Haunted
Empire

Haunted Empire

Apple After Steve Jobs

Yukari Iwatani Kane

HARPER
BUSINESS

An Imprint of HarperCollins*Publishers*
www.harpercollins.com

HarperCollins books may be purchased for educational, business, or sales promotional use. For information, please e-mail the Special Markets Department at SPsales@harpercollins.com.

FIRST EDITION

Designed by Leah Carlson-Stanisic

Library of Congress Cataloging-in-Publication Data has been applied for.

ISBN 978-0-06-212825-6
ISBN 978-0-06-232683-6 (International Edition)

14 15 16 17 18   OV/RRD   10 9 8 7 6 5 4 3 2 1

For Patrick

Contents

Author's Note

This is a work of nonfiction, based on five years of reporting, including three years that I spent covering Apple for the *Wall Street Journal*. All of the names and details described are real.

When I embarked on this project before Steve Jobs's death, I had planned to chronicle how the CEO and his team had rescued Apple from near bankruptcy and turned it into a breathtakingly successful empire. About a year into my reporting, however, I realized that a more compelling story about the company's leadership transition was unfolding right in front of me. Having previously covered Sony and seen its decline following the departure of its founder, I was particularly interested in how Apple would handle these challenging first years in an increasingly complex business environment. And so I started over again with one question—can a great company stay great without its visionary leader? I thought that if any company could, it would be Apple.

Although I had access to the company's media events and some of its executives during my reporting for the newspaper, Apple chose not to grant any further access apart from one shareholders' meeting. Even so, I was able to draw from more than two hundred interviews with nearly two hundred sources who have firsthand knowledge of Apple's world in the United States, Europe, and Asia. They include Apple executives and employees—past and present—as well as business partners, lawyers, friends, and acquaintances who have come into close contact with Apple's inner circle at various points in the company's history. I also interviewed former Foxconn and Samsung employees, executives, consultants, and business partners. Because of the secretive nature of all three of these corporate giants, most of these sources asked not to be named for fear of repercussions. A couple of sources feared reprisals from the Chinese government.

In pursuit of this story, I traveled around the world, starting with the company's headquarters in Cupertino, California, before heading for Chicago, Boston, London, Frankfurt, Beijing, Hong Kong, Seoul, and Tokyo. I visited Tim Cook's hometown of Robertsdale, Alabama, interviewing the CEO's former teachers, driving by his high school, and having fried chicken at Mama Lou's restaurant. I attended the *Apple v. Samsung* trial in San Jose, visited a black market in Shenzhen, and watched the waves of workers entering the gates of Foxconn's massive complex in Longhua. In Taipei, I went to Hon Hai's headquarters in the industrial Tucheng district, where stern-looking guards forbade me from taking photos of the building from their side of the road. I also scoured public records of Apple's corporate dealings and reviewed thousands of pages of court transcripts, internal memos, company emails, and other documents, all of which helped me piece together parts of the story.

Though I witnessed some of the scenes and dialogue in the book firsthand or watched them on videos, other sections are reconstructed from interviews, transcripts, and research. By necessity, some details are based on the recollections of my sources. Mindful of the vagaries of memory, I have made every attempt to confirm their accuracy. When I mention someone by name, readers should not assume that the subject granted me an interview. Many of the statements or occurrences unfolded before an audience or became widely known quickly as they were shared inside Apple.

In many sections, I consulted experts in various fields for help in providing background and context on technical subjects such as patent law, corporate governance, and software design. I also drew on the insights, observations, and reporting by esteemed journalist colleagues around the world, who generously provided material to supplement my reporting and are mentioned by name in the acknowledgments.

To tell this truly global story, I relied extensively on news articles and books written not just in English but also in Chinese, Korean, and Japanese. Having been partly educated in Japan, I had no trouble reading the Japanese articles. For the Chinese and Korean materials, I relied on assistants fluent in those languages.

For specifics on how each chapter was reported, please see the endnotes.

I Used to Rule the World

That Wednesday, the empire went silent.

Across the country, from Boston to San Francisco, Apple stores shooed away customers in the middle of the day and locked their doors. In Chicago, the staff hung a white curtain across its glass storefront. In Washington, D.C., a security guard stood watch in front of the entrance. In Manhattan, the lights and computers were still on, but the floor was eerily vacant.

Inside all of these stores, employees gathered around video monitors for the start of the memorial service to honor the untimely death of their visionary leader. Steve Jobs had been battling cancer for years, so his passing in early October 2011 had not been surprising, but it was no less devastating. In Apple's Tokyo store, employees openly cried. It was the middle of the night, but they had come in just for this occasion. They had been present when Jobs had stopped by a few years before, and it was inconceivable for them not to bear witness to his last gathering.

On the other side of the world, at Apple's One Infinite Loop headquarters in Northern California, it was morning. Fans from near and far had made a pilgrimage to the campus, placing flowers, balloons, and notes in a makeshift memorial alongside the sidewalk in front of Jobs's office building. As employees headed to the courtyard for the ceremony, they passed a colorful string of a thousand paper cranes hanging on a tree in a Japanese symbol of peace. The American, Californian, and Apple flags at the entrance flew at half-staff. Posted signs asked employees to refrain from putting up photos online. Secrecy was law at Apple, but it was particularly important on this day. The company wanted to mourn the loss of its CEO quietly, away from the public's gaze.

One enterprising television station dispatched a helicopter that hovered over the campus with a video camera that captured the scene. The live footage showed people packed around the company's outdoor amphitheater. Fall was coming, and the leaves on the trees were blushing red. Thousands of employees filled the courtyard. More lined up outside as shuttle buses delivered workers from Apple's satellite offices. Jobs's widow, Laurene, sat discreetly to the left of the stage. Dressed in black, her eyes hidden behind sunglasses, she flashed the barest of smiles.

Apple employees whose offices faced the courtyard looked out from their balconies. Next to them, draped on the buildings, were massive black-and-white photos of Jobs two stories high. The deification of the fallen emperor had begun. In one photo, a young Jobs sat in lotus position, cradling an original Macintosh computer in his lap. In another shot from 2004, Jobs clasped his hands, a hint of a smile suggesting a quiet confidence, almost as if he foresaw Apple's coming ascendency. The third image would adorn the cover of his biography—a bearded Jobs with his hand touching his chin—a portrait of a man who knew he had changed the world.

Nearby lay stacks of white program books with the title "Remembering Steve." Inside was a copy of a commencement speech that Jobs had given at Stanford University in 2005.

"Your time is limited, so don't waste it living someone else's life," it said. "Don't be trapped by dogma—which is living with the results of other people's thinking. Don't let the noise of others' opinions drown out your own inner voice. And most important, have the courage to follow your heart and intuition."

Lost in the inspirational messages was the hubristic side of Jobs: arrogant and controlling, an obsessive-compulsive tyrant. Now that he was gone, these complexities only added to his myth.

Although Apple's faithful were gathered to celebrate their CEO's extraordinary life, many in attendance were eager to prove that Apple would endure without him. The executive team had been running the company for some time, but they were painfully conscious of the immense challenge ahead. The world would be watching for any sign of faltering. Former vice president Al Gore, a member of Apple's board of directors, told the crowd to have faith. Jobs had prepared them for this moment, instilling the passion and drive to dream up transformative products. "Keep on skating to where the puck will be," he said, repeating the Wayne Gretzky line that Jobs used to quote.

The crowd was comforted when Jonathan Ive, Apple's lead industrial designer, appeared onstage. He and Jobs had created Apple's beautifully designed products. He was Jobs's closest colleague.

"He, better than anyone, understood that while ideas ultimately can be so powerful, they begin as fragile, barely formed thoughts so easily missed, so easily compromised, so easily just squished," Ive said. "His, I think, was a victory for beauty, for purity, and . . . for giving a damn."

Now it would be up to Ive to keep those ideals from slipping away. The same was true of Jobs's other lieutenants. Each shouldered the responsibility for Apple's continued success.

Tim Cook, the company's new CEO, was known as a stoic. But as he stepped to the microphone to talk about Apple's loss, his voice cracked.

"The last two weeks for me have been the saddest of my life by far," Cook admitted. Jobs, he said, had been called "a visionary, a creative genius, a rebel, a nonconformist, an original, the greatest CEO ever, the best innovator of all time, the ultimate entrepreneur. He had the curiosity of a child and the mind of a genius. All of these are true and the fact that all of them were embodied in one man is amazing. But for those of us who knew and loved him, none of these words, by themselves or in total, adequately define who Steve was."

Cook summarized Jobs's thinking eloquently, quoting some of Jobs's most famous credos.

"Simple can be harder than complex. . . . You have to work hard to get your thinking clean, to make it simple. . . . Just figure out what's next."

Knowing what to do next was one thing. Executing it was another. Apple's business in the past few years had become much more complex. Bigger, more global, and higher profile, the company now had much to lose. Apple was engaged in a fierce global battle against rivals in the smartphone and tablet markets, and it was under greater scrutiny from the government. Cook had to manage a sprawling supply chain in Asia, while also satisfying the public's insatiable appetite for "insanely great" products. Jobs's stupendous feats had built their expectations to stratospheric levels, and each success made the next one that much harder to achieve.

Jobs didn't expect Cook to do what he would have done. He didn't even want Cook to ask that question.

"Just do what's right," he had advised.

After working side by side with Jobs for almost fifteen years, Cook found his absence inconceivable. How would Cook make his mark in a company so infused with Jobs's persona that even the bottle of water placed at his side was Glacéau Smartwater—Jobs's brand of choice? Apple was Jobs and Jobs was Apple. Throughout history, many an empire had fallen into chaos or faded into irrelevance after the death of a beloved but feared leader. How would Cook steer Apple away from the same fate?

At the end of his remarks, Cook played a little-known audio recording of the words to Apple's famous "Think Different" campaign as read by Jobs. His disembodied voice played over the reverent crowd.

"Here's to the crazy ones, the misfits, the rebels, the troublemakers, the round pegs in the square holes, the ones who see things differently."

The words quietly sank in. No wonder Jobs had identified with this ad. It described him precisely.

"They're not fond of rules and they have no respect for the status quo."

He'd subverted not just rules but the truth. He'd created his own reality.

"You can quote them, disagree with them, glorify or vilify them. About the only thing you can't do is ignore them."

The ceremony ended with a performance by Coldplay, one of Jobs's favorite bands. The lyrics to "Viva la Vida" floated across the courtyard.

I used to rule the world
Seas would rise when I gave the word

Now in the morning I sleep alone
Sweep the streets I used to own
I used to roll the dice
Feel the fear in my enemy's eyes
Listen as the crowd would sing
"Now the old king is dead! Long live the king!"

As the throngs departed, their emperor watched on, gazing down upon them from the banners. Unblinking. Looming. Waiting to see how his lieutenants would shepherd his creation into the future.

The Disappearing Visionary

JUNE 2008

For a company at the epicenter of American business and culture, Apple's headquarters couldn't have been located in a more forgettable place. An hour's drive south of San Francisco, the campus was situated deep in Silicon Valley in Cupertino, a quiet suburb where technology companies coexisted with relatively modest residential homes, big-box retailers, and chain restaurants like T.G.I. Friday's. As cities go, Cupertino was nondescript. It didn't have a downtown to speak of, nor did it have a major shopping mall. Many of Apple's employees preferred to live in San Francisco, so the company ran luxury commuter buses for them with wireless Internet and leather seats. Every weekday morning, the buses rolled down Interstate 280 alongside the Santa Cruz Mountains to arrive at De Anza Boulevard. Occasionally, Jobs would be spotted cutting off other drivers as he turned onto the campus road in his silver Mercedes-Benz convertible. His car stood out because it had no license plate.

Apple's many offices were sprinkled throughout the area. The iTunes team inhabited buildings on Valley Green Drive. The marketing and communications department worked out of a large office on Mariani Avenue. Passersby would not know that a building belonged to Apple unless they happened to spot the discreet sign with a small Apple logo. That is, if it had a sign at all.

It was only after visitors turned into a side street to the east of De Anza behind BJ's Restaurant and Brewhouse that they could see a group of six buildings that together formed Apple's main oval-shaped campus. It was Cupertino's most famous address: One Infinite Loop. The four-story buildings, labeled IL1 through 6, looked mostly the same from the outside—concrete boxes with plenty of windows that gave them a clean, open feel. But the interiors were designed by different architects to give each building its own personality. IL2, which housed the iPhone software team, had a 1990s-style postmodern look with angles and curves. The industrial design studio, which shared part of the building, had a sleek interior with frosted glass and stainless steel. IL6, where the operations team resided, was more classic and subdued. The names of meeting rooms in each of the buildings reflected Apple's playful culture. The developer relations team in IL3 named their rooms after evangelists like Tammy Faye and Pat Robertson. The product marketing team called theirs "Here," "There," and "North by Northwest." The iPhone software team's rooms were tongue-in-cheek. "Between" was literally flanked by two rooms called "Rock" and "A Hard Place."

What the buildings had in common was that they forced people to interact. Offices were lined with windows. Hallways opened up into common spaces. The floor plan made it impossible not to bump into other teams. Longtime Apple employees swore that these informal encounters were part of the secret to its success because they fostered collaboration.

The first building that one saw when driving onto campus was IL1, also known as "Steve's building." Part of the ground floor was occupied by the company store, where Apple devotees could buy T-shirts that said, "I visited the mothership." If they looked past the lime-green sign with the number "1" and the manicured evergreen hedge out front, they would see employees hanging out in an airy, glass atrium. The executive offices were on the top floor, but access was restricted. Jobs occupied a corner office facing the quad. It contained a huge desk, a couple of chairs, a sofa, coffee table, and a credenza. Piles of books, papers, and random items that people sent to him were strewn everywhere. It was so functional and devoid of character that executives who had spent time there struggled to describe it.

Jobs didn't care what his office looked like. He was hardly ever there. He met most people in the boardroom by the floor entrance or in the conference room next to his office. Otherwise, he'd usually be walking around, hanging out with Ive, or visiting the industrial design studio, where there were always interesting projects and prototypes to look at. Jobs loved the studio. It was one of the reasons he had it moved some years ago from the other side of De Anza Boulevard to IL2, right next door to him.

One morning that summer, Jobs was sitting in the front row of Town Hall—Apple's auditorium—doing one of the things he did best: terrifying others into outperforming themselves. The Worldwide Developers Conference, Apple's annual meeting for developers, was coming up soon, and he was there to watch the rehearsals.

Every WWDC was important, but the 2008 meeting was particularly so because Apple was about to launch the App Store. The company needed to get developers excited about submitting apps. More than five thousand people would be attending, and many more would be following the event through blogs and media reports. Jobs needed to win over every one of them. To help him, he was inviting a few developers onstage during his keynote address to showcase their apps.

First, however, Jobs was vetting their presentations personally. He wasn't about to allow Apple's carefully crafted image to be bruised by technical glitches or a developer who stumbled over his words. The developer relations team had spent the last several weeks choosing the final candidates. After they had spent a couple of days refining their apps and polishing their two-minute demo scripts behind closed doors at Apple's offices, the time had come for the developers to show Jobs what they had. The speakers had said their lines dozens of times, but this would be their biggest test yet.

Everyone knew Jobs was a perfectionist with a fiery temper. Making a presentation in front of him was a challenge, even for veterans accustomed to his acid tongue. "What the fuck is that?" Jobs would ask when he saw something he didn't like. It was rare for someone to actually reach the end of a presentation without Jobs jumping ahead to the conclusion or going into a tirade. "Your communication is so poor that I can't even tell what you're talking about," he once told someone who was giving him a private demonstration. "Until you guys figure out how to communicate, I don't know how we can even have a discussion."

As Jobs's intense brown eyes bore down on them, speakers stood onstage to practice their lines. The wait to rehearse in front of Jobs was torturous. One of the staff members tried to put the presenters at ease. "Don't worry," he said. "Steve's a normal guy. He puts his pants on one leg at a time."

It didn't reassure anyone. A mistake at any point could cost the presenters their slot. When their turn came, speakers entered the auditorium and walked down the aisle, passing by Jobs. The air crackled with tension.

"Is yellow eBay's corporate color?" he asked after its speaker finished his spiel. He didn't like the color in the app, but when the answer was yes, he backed off. Occasionally, he'd provide small suggestions directly to the speakers, asking them to emphasize a particular point or change something on their app.

Most of them were prepared, thanks to the intense coaching from Apple's team. Jobs, in a benevolent mood, smiled and told them they'd done a great job.

As the teams finished, they felt the weight slide off their shoulders. There would be two more dress rehearsals before they would make it onstage, but they had survived an encounter with Jobs, and they were elated.

That day, the developers were too focused on themselves to notice that the CEO was looking frail and gaunt.

Only fifty-three, Apple's savior had already conquered much of the modern world—transcending the computer industry and redefining industries as varied as retail, music, and mobile phones. It was hard to believe he was mortal. Even in his daily life, he defied the natural order. The fact that he drove without a license plate demonstrated how he soared above the everyday concerns of his fellow humans. He got away with it by leasing an identical car every six months, within the grace period California state laws set to obtain plates for new cars.

Apple would never have been created or resurrected without Jobs, and it was difficult to imagine the place continuing to thrive without him. Through the years, the company's hopes and aspirations had become completely intertwined with his continued success.

Apple had been officially founded on April Fools' Day in 1976. A twenty-one-year-old Steve Jobs and his twenty-five-year-old friend Steve Wozniak started their computer business in the garage of Jobs's parents' home in Los Altos, California.

Jobs had a vision of computers as mental bicycles, a tool that helped make the most of a user's intellect in the same way that bicycles helped make the most of a rider's athletic ability. But the board of directors considered him too inexperienced to manage the company and asked him to hire a CEO. When he found John Sculley, a former PepsiCo president, he recruited him with the famous words:

"Do you really want to sell sugar water for the rest of your life or do you want to come with me and change the world?"

The following year, Apple launched the original Macintosh. An infamous sixty-second television spot on Super Bowl Sunday declared the company's intentions, depicting a female runner bursting into a drab room and hurling a sledgehammer at a huge screen projecting Big Brother. In an Orwellian reference to IBM's dominance, a male voice intoned, "On January twenty-fourth, Apple computer will introduce Macintosh. And you'll see why 1984 won't be like *1984*."

The next week, a proud Jobs stood on the podium at Apple's annual shareholders' meeting in a dark suit and bow tie and pulled its newest computer out of a bag. As the computer turned on and showed all the things it could do, the theme song to *Chariots of Fire* played in the background. Then the machine spoke. "Hello, I am Macintosh. It sure is great to get out of that bag." The audience went wild, and a cult was born.

But that would be Jobs's last moment of glory for a long time. His disruptive behavior and pursuit of perfection at all costs had wreaked havoc on the company, and the executive team wanted him out. A little over a year later, he was in exile.

Under Sculley, Apple flourished for many years as consumers paid a premium for the Mac's unique mouse-based point-and-click user interface. But when Microsoft's Windows software caught up and exceeded the Mac's features, the kingdom crumbled. PCs running Windows soon proliferated, and Apple's market share shrank.

By the mid-1990s, long after Sculley had left and his successor had come and gone, Apple was on the brink of bankruptcy, forced to cut prices and add new product lines to sustain its growth. The company was selling dozens of computer models that were confusingly similar but had incompatible operating systems. Employees began jumping ship. Three days after Gil Amelio, the former CEO of National Semiconductor, took over, *BusinessWeek* published a story with the title "The Fall of an American Icon."

Amelio was in way over his head. Apple needed a leader who could take extraordinary measures at breakneck speed, but Amelio was a classic, hands-off corporate executive, who preferred to supervise rather than take action himself. He hired image consultants, created new acronyms to describe business concepts, and put together white papers. He was also a bad fit culturally. He cared about executive perks and liked formality. He drove a Cadillac Seville and always ate his lunch on china. His executives mockingly called it "Gil's special china." According to rumor, it was Wedgwood. Each day one of his assistants had the unenviable job of taking his meal out of a takeout container and plating it.

For all the criticism about his tenure, Amelio had made one of Apple's most critical moves—the decision in December 1996 to buy NeXT, Steve Jobs's failing computer company. As part of the deal, Jobs became Amelio's advisor.

Amelio's tenure unraveled quickly, accelerated in part by a scornful Jobs, who had plenty to say about Amelio's management decisions behind his back but offered little advice. He attended one executive team meeting, but he walked out in the middle of it and never returned. At a dinner party to celebrate the sale of NeXT, Jobs joked about creating a "Gil-o-meter" to gauge stupidity. "Two Gils" meant someone was being twice as stupid as Amelio.

After a famously disastrous Macworld speech in January, where Amelio was upstaged by Jobs, the CEO lost the board's confidence. During his short reign, Apple had lost more than $1.6 billion. The company was so undesirable that no one even wanted to buy it. Its brand name would maybe fetch $500 million.

After the board begged him to return, Jobs overhauled the company's culture. During his absence, Apple had fallen into complacency as everyone took more interest in celebrating their accomplishments than breaking new ground. There were team T-shirts to commemorate new projects and a garden of Macintosh icon sculptures and a display of an old Apple I to remember the company's past successes. Employees took six-week sabbaticals every five years.

That changed overnight. Jobs ordered the mementos removed and the people on sabbaticals recalled. He forbade hard liquor, smoking, and pets, and replaced Amelio's china with ordinary cafeteria dishes. He did away with anything that he perceived as corporatized and installed a meritocracy that rewarded agility, ambition, and boldness.

"The lunatics have taken over the asylum, and we can do anything we want," Jobs joked shortly after his return.

Accustomed to a more laid-back work environment, many people left the company voluntarily. Others were fired. Stories circulated about Jobs going into meetings and terminating people on the spot. A rumor started that he had fired someone in the elevator. When a protective covering went up in the elevator of IL1 to shield it from construction work, one person quipped: "This must be Steve's elevator since it's padded." To which a colleague asked, "Is it for him or for us?"

At some point someone made up a verb to describe such unfortunate outcomes: Steve'd.

Jobs also clamped down on secrecy. Outgoing email was monitored, and anyone caught sending messages labeled "confidential" received a warning. Employees, who were accustomed to wandering through open doors, were suddenly prohibited from entering many areas. Office windows were mysteriously covered, and engineers were asked to work on projects without knowing what the product was. Everything was on a need-to-know basis.

This was a complete turnabout. Jobs had once leaked so many corporate secrets that a colleague once teased him, "It's a strange ship that leaks from the top." But he had learned the power of mystery. If Apple had any hope of surviving, it needed to stay nimble and ready to modify its strategy without being hampered by public opinion. Product introductions were also more dramatic when no one knew what was coming. The media attention he received for them was worth millions of dollars in free advertising. Jobs loved the moment of revelation when he introduced a new product that no one in the audience had seen before.

In Jobs's first months back, some employees rebelled. A prankster forged Jobs's email address and sent out a memo. "You've all become lazy and only contribute to Apple's current situation," he wrote, according to one account. "You're now going to have to pay for the water in our water fountains and we're going to add a charge that you'll find on your paycheck for the oxygen that you use for your eight hours on the job." He added that employees would be charged three dollars a day for parking. "Only I will be allowed to park in handicapped spaces," it said, making fun of Jobs's well-known habit.

Twenty minutes later, the real Jobs sent out an email.

"I'm all for having fun," it read, "but we need to be focused on the future in making the company a better place. Best, Steve." The culprit was fired.

In September 1997, Jobs finally agreed to update his status from advisor to interim CEO. Though he wouldn't be ready to permanently commit to Apple for another two and a half years, he started the hard work of rebuilding Apple, killing unprofitable projects and ridding the company of deadwood. He surrounded himself with brilliant lieutenants. Two of them had been with him since his NeXT days. Another two had been at Apple since the previous regime but had impressed Jobs with their eagerness to transform the company. Among his new hires was Tim Cook, who was put in charge of streamlining Apple's operations.

Before Apple could put any more products on the market, it needed to brush up its severely tarnished brand image. To do that, Jobs engaged the creators of Apple's "1984" ad at TBWA\Chiat\Day. This time, they came up with "Think Different," which would go down in advertising history as an unparalleled campaign. Instead of showcasing its products, the ad associated Apple with history's geniuses, cementing its identity as an innovator. If the company's vision was murky before, this clarified it: Apple made products for people itching to change the world.

Once the company reestablished its brand and provided a visionary framework for its future products, Jobs pared Apple's product lines from dozens to just four: one set of desktop and laptop computers for professionals and another for consumers.

The gumdrop-shaped, partially translucent iMac led the charge.

Its unique, fashionable design put Apple back on the industry map. In 2001 Apple opened its first Apple Store, creating a completely new electronics retail experience that invited customers to hang out and try its products. The iPod digital music player and iTunes came next.

Each successive hit increased the mystique around Jobs. By this time, he had perfected the art of secrecy and the only facts that the public knew about him were what he wanted them to know. He used that to cultivate an image of himself as a charismatic superstar who answered to no one but himself. Before long, he became the most recognizable figure in business. He wore the same outfit—a black turtleneck, Levi 501 jeans, and New Balance 991 sneakers—until it became his trademark.

But in October 2003, the unthinkable happened. Jobs was diagnosed with pancreatic cancer. The initial report was bleak. He was told that he had three to six months to live, and he should get his affairs in order. A biopsy later revealed that he had a rare kind of treatable cancer, but it was still cancer. As Jobs broke the news to his executive team, he cried.

Jobs considered disclosing his illness to the public but changed his mind. He didn't want people to view him as a helpless patient. Besides, once he disclosed information about his health, it would be irretrievable. The brouhaha that would ensue would be an intrusive distraction. His lawyers determined that it wasn't a problem as long as Jobs was able to serve as CEO.

Jobs made his recovery difficult. Though doctors advised surgery, he ignored them because he didn't want them cutting into his body. He pursued an alternative treatment through a vegan diet, acupuncture, and herbal remedies. "Voodoo medicine," Jobs's friend and board member Bill Campbell called it. A few people who worked with him closely noticed his flagging energy and his unusually eccentric diet of juices and broths. Only his innermost circle knew what was wrong. Those who brought attention to it were told to forget that they had noticed.

By July of the following year, however, Jobs wasn't getting any better, and he underwent surgery at Stanford University Medical Center to remove a part of his pancreas. The following day, he sent out an email to employees. "I had a very rare form of pancreatic cancer called an islet cell neuroendocrine tumor, which represents about 1 percent of the total cases of pancreatic cancer diagnosed each year, and can be cured by surgical removal if diagnosed in time (mine was)."

The email wasn't entirely truthful. Jobs neglected to explain that he had been diagnosed nine months earlier. While it was true that Jobs's kind of cancer might have been cured by timely surgery, he had waited too long. During the operation, doctors found three liver metastases. But that fact was kept quiet, and Jobs's protectors closed ranks. Campbell told a reporter that the board didn't think it was necessary to disclose the timing of his diagnosis because Jobs had been working right up until his surgery. A question about Jobs's sick appearance at WWDC one year was laughed off.

"I wish someone would comment like this when I lose weight," Campbell joked. An Apple spokeswoman pointed out, "He's six feet tall and weighs 165 pounds. It's a healthy weight for his height."

Jobs supported the storyline, telling everyone that he had been "cured." In June of the following year, he gave the commencement speech at Stanford University that addressed his mortality. "Remembering that I'll be dead soon is the most important tool I've ever encountered to help me make the big choices in life," he said. "Remembering that you are going to die is the best way I know to avoid the trap of thinking you have something to lose. You are already naked. There is no reason not to follow your heart."

Jobs began working even more feverishly. After two years of intense development, Apple launched the iPhone in 2007. Unlike the typical cell phones of the time, it was a sleek and elegant phone with a large rectangular touchscreen and a button. Users could check email, surf the Web, play music, and perform many other functions with the touch of a virtual button or a swipe of a finger. A simple interface made it delightfully accessible to even the most technology-shy consumers. As with its other products, the iPhone pushed the envelope in design. Some dubbed it "the Jesus Phone."

Apple made history again. There was much excitement in the company about how it could follow this promising debut. The App Store was just one of them.

But once again, Jobs's health was interfering.

The cancer was affecting his body, and he was in increasing pain. Morphine helped, but it reduced his appetite and added to digestive problems stemming from his first surgery. He had trouble keeping food down. He told one friend that "when he feels really bad, he just concentrates on the pain, goes into the pain, and that seems to dissipate it." That spring, while the company prepared for WWDC, Jobs had lost forty pounds.

His executives watched him wither even as they pretended that everything was okay.

In the week before the developers conference, Jobs was running a high fever. But the focus of discussion among the planning staff was on his appearance. The CEO's keynote was a marquee event at WWDC, and everyone would be paying close attention. It was unlikely that the audience would overlook his weight loss. As a solution, someone proposed that Jobs wear two turtlenecks. The idea was discarded in part because no one dared to suggest it to him.

By 7 a.m. on the first day of WWDC, the line already stretched around the corner of the Moscone Center in San Francisco. Jobs's keynote wasn't due to start for another three hours, but his disciples arrived early in the hope of getting a prime seat. Many had traveled from around the world. The long-distance travelers stood out because they weren't dressed for the Bay Area's chilly summer. While the locals came prepared in long-sleeved shirts or jackets, the out-of-towners shivered in short-sleeved shirts and jeans or shorts. A classic tourist mistake.

A few minutes before 9 a.m., there was a flurry of excitement as some of them saw that the online Apple store had posted a yellow "We'll be back soon" sign. That meant they were queuing up pages with information about the new products they would unveil in the next few hours.

A short while later, Apple opened the doors to the auditorium, where Bo Diddley and Chuck Berry blared through the speakers. Reporters craned their necks to see if they could spot any executives in the front of the room. As they eagerly reported sightings of Chief Operating Officer Tim Cook, marketing chief Phil Schiller, and board member Al Gore on their live blogs, the lights began to dim and the last song, Jerry Lee Lewis's "Great Balls of Fire," ended.

When Jobs appeared from the left side of the stage, the audience broke into thunderous applause as cameras flashed around the room. Some whistled in appreciation as a grinning Jobs soaked it all in before talking up the brilliance of Apple's products and services. He was so persuasive that he seemed to be casting a spell. The fans called it his reality distortion field. The developers, who had been chosen to demo their apps, performed their roles perfectly.

"Isn't that fantastic!" he enthused. "This is going to be great."

With a half hour left in the presentation, Jobs began talking about the new iPhone, with long-awaited features like 3G high-speed wireless connectivity and third-party apps. The starting price: $199 compared to the $399 price for the first phone.

Jobs ended his performance with a showing of Apple's latest ad, which poked fun at Apple's secrecy. In the television spot, two guards dressed in dark clothing carry a metal box through an austere corridor as they slide their security cards to enter a protected area under the watch of a security camera. There, they take their key to unlock the box, which reveals the iPhone 3G.

"Isn't that nice?" Jobs asked when it was over. "You wanna see that again? Let's roll that again. I love this ad!"

It was another masterful performance, but the media soon turned their attention to Jobs's dramatic weight loss. Reporters who interviewed Jobs afterward noticed that his collarbone was visible through his shirt. Jobs's strange dietary habits were well known, and his weight had fluctuated before, but this time he was looking emaciated. Gossip site Gawker was one of the first to speculate publicly about whether the cancer had come back.

"People watching the imperiously slim presenter at the WWDC today are finding it hard to look at Jobs's frailer than ever frame and not wonder if he's still suffering," the site wrote.

When the *Wall Street Journal* contacted the company for comment about Jobs's extreme weight loss, Katie Cotton, Apple's spokeswoman, tried to neutralize the story, saying that Jobs had picked up a "common bug" in the weeks before the event and had been taking antibiotics. The Internet exploded when the *Drudge Report* saw the ensuing *Journal* piece and linked to it. Many bloggers and journalists rose to the company's defense, saying that Apple would have issued a release if Jobs were really ill, but others demanded more information.

"For almost any other human being, this topic would be a personal matter," wrote Henry Blodget, a well-known equities analyst from the dot-com era, in a blog entry. "In this case, however, tens of billions of dollars of market value rests on Steve's remaining healthy and at the helm of Apple for many years, so his health is a material business concern."

The world was waking up to the fact that the emperor was wasting away.

2

Reality Distortion

One Thursday afternoon in late July a few weeks after WWDC, the *New York Times* columnist Joe Nocera was working at his desk when his phone rang. He didn't recognize the number, but the 408 area code told him that it was from someone in Silicon Valley. Curious, he picked up and heard the voice of someone he hadn't spoken with in decades.

The first words out of the caller's mouth made Nocera sit up.

"This is Steve Jobs. You think I'm an arrogant asshole who thinks he's above the law, and I think you're a slime bucket who gets most of his facts wrong."

Nocera had been working on a column about Apple's lack of disclosure on Jobs's health, and he had asked the company spokesman for a comment. He hadn't expected to hear back from Jobs himself.

In the weeks since the WWDC keynote address, the speculation over Jobs's weight loss had continued, but Apple had said little. In its quarterly earnings call, Chief Financial Officer Peter Oppenheimer deflected the questions about Jobs's health with a vague nondenial.

"Steve loves Apple. He serves as the CEO at the pleasure of Apple's board and has no plans to leave Apple. Steve's health is a private matter."

The company didn't comment at all when the *New York Times* published a story about Jobs's health. The article revealed that Jobs was dealing with nutritional problems after his 2004 cancer surgery and had had another surgical procedure earlier in the year related to his weight loss, but that he had been reassuring people around him that he was cancer-free. It was mystifying that Apple wouldn't want to endorse this story in some way if it were true. It would have calmed the fears that had caused Apple's stock to fall from $186 before WWDC to as low as $149 within about six weeks.

That's when Jobs came calling. Nocera hadn't talked to him in twenty years. The last time was in 1986, when he worked on an article for *Esquire* and shadowed Jobs for a week shortly after he had started NeXT. Nocera had gotten unimaginable access to him back then, sitting in on meetings, dining together at one of Jobs's favorite vegetarian restaurants, and even driving with him to Pixar. But he had never heard from Jobs afterward and hadn't maintained the relationship.

"Do you want to have a conversation about this off the record or not?" Jobs asked.

Nocera wanted it on the record, but Jobs refused. He wasn't really asking. This was another ploy that he used often. Off-record conversations allowed him to influence articles without putting himself on the line. If Nocera wanted to hear what Jobs had to say, he had no choice but to acquiesce.

In the remarkable conversation that followed, Jobs wove a dubious story. Echoing the company line, Jobs first told Nocera that he had had a major infection and a high fever before WWDC and that he had been taking antibiotics.

"It wasn't cancer related in the slightest," he promised. He acknowledged that he had undergone a small, outpatient procedure several weeks before, but insisted that calling it "surgery" was inaccurate. "It's like having a wart removed," he said, his tone measured and calm. He repeated that it had nothing to do with cancer.

Nocera pressed his central question—why wasn't Apple saying more about his health?

Jobs grew defensive. He reminded Nocera that he had immediately disclosed his surgery back in 2004. In comparison, former Intel CEO Andrew Grove had waited a year after his operation to tell the public about his prostate cancer.

"You're saying I should have disclosed it the day before, but there's no clear policy," Jobs said, adding speciously that more people died from prostate cancer than from what he had. He then asserted that his obligation to provide details about his health was somewhere between that of a janitor and the president of the United States.

Jobs had been feeling as if he were under attack from the media. He was particularly incensed over a *Fortune* magazine article that claimed his surgery a few years before was called the "Whipple procedure," in which parts of the pancreas, stomach, small intestine, and bile duct were removed. It wasn't Nocera's story, but Jobs wanted to set the record straight.

"It was a semi-Whipple procedure. I had roughly half of my pancreas removed. I did not have my spleen removed," he said, splitting hairs. What he described was just a modified version used to preserve the stomach and minimize potential nutritional problems. Then he circled back to his previous point. His health was his business.

"Someday I won't be CEO of Apple, and it'll be time to have some new blood in there, so why don't we pretend that today? Anybody who doesn't want to be at Apple without me should just sell their stock."

Nocera just sat back and listened. He was mildly tongue-tied and somewhat abashed by the outpouring of emotion. Jobs wasn't looking for a dialogue. He just wanted to vent.

"Don't get on your soapbox," Jobs snapped. He clearly knew Nocera's position on the issue. Nocera felt that Jobs had an obligation to be transparent on matters as crucial as this. He was arguably the most indispensable chief executive on the planet. And he had had cancer.

"Don't use me as a whipping boy on this. My private life does not belong to the shareholders. If they want me to step down, the directors should remove me from the job," he said, adding that otherwise they should "shut up."

"Where is everybody demanding that Rupert Murdoch publish his cholesterol level?"

Unless there was a change in policy, Jobs said, the public needed to accept that he wasn't going to talk about his health.

"I don't begrudge you your belief, but I have a different belief. I don't hold myself above other CEOs as you hold me."

He wrapped up the call by pointing out that he wasn't the only executive at Apple. "I think we have great leaders at Apple. The last few public events, I've tried to feature those other leaders. I've been trying to expose the other people."

The conversation was classic Steve. Slices of what he said may have been technically true. Taken together, it was wildly misleading. Jobs's deflection tactics were famous, but not being privy to the details about the CEO's health, Nocera had no basis to doubt him. In any case, other than the "slime bucket" line, everything about the phone call was off the record, so he couldn't quote Jobs on any of it.

In his actual column, Nocera stuck to his criticism of the company, writing, "Apple simply can't be trusted to tell the truth about its chief executive." But he also provided the reassurance that investors had been looking for.

"While his health problems amounted to a good deal more than 'a common bug,'" he wrote, "they weren't life-threatening and he doesn't have a recurrence of cancer." In the end, Jobs got what he wanted. The speculation that plagued the company died down. Apple's stock eventually recovered to around $180.

Once the idea that Jobs was ailing had been planted in the public's mind, however, it was impossible to expunge it. Reporters kept digging and pre-wrote obituaries. In late August, Bloomberg inadvertently released its draft. "Steve Jobs, who helped make personal computers as easy to use as telephones, changed the way animated films are made, persuaded consumers to tune into digital music and refashioned the mobile phone, has XXXX. He was TK," the draft said. "XXXX" was a placeholder for the details of his death, and "TK" was a copyediting code indicating that there was more information to come. At the top was a list of Jobs's friends and colleagues whom the news service planned to contact for a comment. Jobs seemed to take it in stride, at least publicly. When Apple held a press event shortly after, Jobs put up a slide with Mark Twain's quote "The reports of my death are greatly exaggerated."

Beneath the bravado, paranoia lingered. When an Apple contractor posted on the Internet a video of Jobs walking into the office, he was summarily fired. Any speculation concerning Jobs's health was still verboten. Apple's shares sank as low as $85 amid the uncertainty.

At a media event to unveil a new lineup of laptop computers in mid-October, reporters took note right away that Jobs remained thin. Jobs continued to pretend that nothing was wrong. Before a question-

and-answer session, he showed another slide that said "110/70 Steve's blood pressure."

"This is all we're going to talk about Steve's health today," he said.

Just as he had told Nocera, Jobs shared the stage that day with more of his executives than in the past. Tim Cook, Jobs's second-in-command, kicked the morning off with a briefing on the state of the Macintosh computer business. Though he was unaccustomed to the limelight and nowhere nearly as exciting as Jobs, the audience was charmed by his relaxed southern accent. Apple's design superstar Jonathan Ive also made a rare personal appearance. The Brit, who typically participated through videos and remote demonstrations, spoke haltingly but earnestly about the revolutionary design of the latest MacBook Pros. In the question-and-answer session, Jobs and Cook were joined by Schiller, a well-known gadget lover who knew the ins and outs of practically every product feature.

It wasn't lost on the audience that Jobs had played a smaller role than usual. Reporters would look back on the event later and wonder why. Was it a deliberate strategy to show the depth of Apple's executive bench or had Jobs been too sick to manage the presentation?

Several weeks later, it became even clearer that something was wrong. Apple announced that Jobs would not be speaking in January at the Macworld trade show, where he had presented the keynote every year since 1997 and where major products like the MacBook Air laptop and the iPhone were first shown. Apple's explanation was that it was moving away from industry events in favor of its own press events, but few believed this official story. The decision seemed too last-minute.

Renewed fears about Jobs's health swept through the market, and once again Wall Street was flooded with phone calls from investors. Apple's shares lost 6.6 percent of their value by the end of the following day.

Apple tried to quell the speculation. "If Steve or the board decides that Steve is no longer capable of doing his job as CEO of Apple," a spokesman told reporters, "I am sure they will let you know."

The truth, no matter how skillfully Apple tried to mask it, was that Jobs was getting sicker. His digestive problems, not uncommon in patients who had the Whipple procedure, were the least of his concerns. The tumors in his liver were messing with his hormones, and drug therapies were giving him uncomfortable side effects. One therapy made his skin dry out and crack, while an experimental treatment involving a radioactive substance caused nausea, vomiting, and abdominal pain. Jobs's best hope for recovery was a liver transplant, but he refused to admit that he needed one.

All of Apple's efforts to protect the CEO's privacy came crumbling down on January 5, 2009, when he sent out an open letter.

Dear Apple Community,

For the first time in a decade, I'm getting to spend the holiday season with my family, rather than intensely preparing for a Macworld keynote.

Unfortunately, my decision to have Phil deliver the Macworld keynote set off another flurry of rumors about my health, with some even publishing stories of me on my deathbed.

I've decided to share something very personal with the Apple community so that we can all relax and enjoy the show tomorrow.

As many of you know, I have been losing weight throughout 2008. The reason has been a mystery to me and my doctors. A few weeks ago, I decided that getting to the root cause of this and reversing it needed to become my #1 priority.

Fortunately, after further testing, my doctors think they have found the cause—a hormone imbalance that has been "robbing" me of the proteins my body needs to be healthy. Sophisticated blood tests have confirmed this diagnosis.

The remedy for this nutritional problem is relatively simple and straightforward, and I've already begun treatment. But, just like I didn't lose this much weight and body mass in a week or a month, my doctors expect it will take me until late this Spring to regain it. I will continue as Apple's CEO during my recovery.

I have given more than my all to Apple for the past 11 years now. I will be the first one to step up and tell our Board of Di-

rectors if I can no longer continue to fulfill my duties as Apple's CEO. I hope the Apple community will support me in my recovery and know that I will always put what is best for Apple first.

So now I've said more than I wanted to say, and all that I am going to say, about this.

Steve

Jobs had been furious about all the speculation over his health following the Macworld announcement. He wrote the letter because he thought Apple's board and staff weren't doing enough to kill the rumors. He was in complete denial that he was gravely ill.

"You guys don't know what you're talking about," he said, lashing out at people. "I'm not that sick. Why aren't you supporting me?"

The company, in fact, had been doing everything they could to protect his privacy. Securities laws didn't define a company's obligation to disclose details about its chief executive's health, but the one clear rule was that a company could not mislead shareholders, and that once it made a statement, it had to update the information as it changed. Apple so far had been extremely careful about crafting its comments so they were technically true. It also had deliberately avoided answering any questions about Jobs's cancer, so they didn't have to update his status. No one would ever hold up their handling of the situation as a model for corporate governance, but it was a compromise they had made to honor Jobs's wish for privacy. It was a delicate balance.

Jobs's letter potentially put Apple into legal hot water. It was true that Jobs had a hormone imbalance. His body was producing an excessive quantity of a hormone called glucagon, which increased blood sugar levels, interfered with his body's ability to use the blood sugar for fuel, and led to a condition in which his body was breaking down fat to use for energy instead. But then Jobs crossed the very line that Apple's lawyers had been trying so hard to avoid. He deceived shareholders by claiming the treatment was "simple and straightforward." Given that the root of the cause was a recurrence of cancer, it was patently untrue.

The letter also undid Apple's efforts to avoid any obligation to keep the public informed of Jobs's health. Jobs had ensnared himself and the company in a trap. The only way out for him to stay compliant with the law was to go on medical leave, which was something he probably should have already done.

After increasingly insistent legal advice, Jobs sent out another letter nine days later:

> *Team,*
>
> *I am sure all of you saw my letter last week sharing something very personal with the Apple community. Unfortunately, the curiosity over my personal health continues to be a distraction not only for me and my family, but everyone else at Apple as well. In addition, during the past week I have learned that my health-related issues are more complex than I originally thought.*
>
> *In order to take myself out of the limelight and focus on my health, and to allow everyone at Apple to focus on delivering extraordinary products, I have decided to take a medical leave of absence until the end of June.*
>
> *I have asked Tim Cook to be responsible for Apple's day to day operations, and I know he and the rest of the executive management team will do a great job. As CEO, I plan to remain involved in major strategic decisions while I am out. Our board of directors fully supports this plan.*
>
> *I look forward to seeing all of you this summer.*
>
> *Steve*

The letter was as disingenuous as the first. Media speculation had nothing to do with why Jobs was taking medical leave. But it brought the public up to date enough on his health and then took him out of the equation so he would no longer be legally obligated to say more.

Reporters and shareholders dissected the two missives, trying to figure out the truth. What had Jobs meant by a "hormone imbalance"? How much was Apple keeping to itself? The company was secretive about everything, but it was usually good about setting the record straight on false rumors. Why wasn't it doing so this time?

Apple wanted the public to believe that there was nothing seriously wrong with Jobs, but they seemed unwilling to make any definitive statements that could reassure the public. Apple's share price dropped below eighty dollars as the doubts crept back about whether Jobs had really just had "a common bug" during WWDC.

Investors began clamoring for information about Apple's succession plan. "This is the biggest wake-up call for the investor community to come to the reality that there will be a post–Steve Jobs Apple," said Gene Munster, an analyst with Piper Jaffray. A few days later, the Securities and Exchange Commission, which regulates the nation's stock markets, opened an informal inquiry into Apple's disclosures.

Employees inside Apple tracked the media reports with dismay as they worried about both Jobs's and Apple's future. Looking back, there had been rumors among some of the managers that Jobs was ill. And there had been times that Jobs seemed more distracted and impatient than usual, flying off the handle for no good reason. But he had always seemed to come around again, and they had bought his story that he wasn't sick. Many employees blamed the media for making an overly big deal out of it.

The split in opinions, however, wasn't only taking place in the public and among employees. It also existed on the board of directors.

Though small, Apple's seven-member board, which included Jobs, was one of the most involved corporate boards around. Mickey Drexler, the chief executive of J.Crew and former head of Gap, was instrumental in setting up Apple retail stores. As chair of the audit committee, Jerry York, the former chief financial officer of IBM and Chrysler, was Apple's watchdog on corporate governance. Former vice president Al Gore applied his political skills to synthesize multiple views and package them into one idea. He was indispensable in creating consensus on the board. Andrea Jung, the chief executive of Avon, was the newest member and was still finding her footing, but the rest of them had been together for a long time, and there was a collegial atmosphere. Even during the Macworld uproar in December, Drexler had stopped by the J.Crew store at the Stanford Shopping Center to pick up a box of Christmas-themed boxers and socks for everyone. The directors were so delighted you would have thought they had never seen underwear before.

Among the board members, its two co-lead directors, Art Levinson and Bill Campbell, had been regularly consulted on Jobs's health. The rest of the board was not as fully informed. Levinson, who had been the chief executive of biotechnology firm Genentech, had extensive medical knowledge that could help Jobs. Genentech also made many of the drugs that Jobs was taking. Campbell, a former Apple executive and Intuit CEO, was a personal friend and confidant who regularly took walks with Jobs. Known affectionately as "Coach" in Silicon Valley for his mentoring of younger entrepreneurs, Campbell was the cheerleader on Apple's board. When Jobs showed off a new product or prototype, Campbell was usually the first one to start clapping. Jobs appreciated his support and tended to listen to him more than to the other board members, so much so that other directors often went to Campbell rather than Jobs when they wanted to bring up an issue.

Levinson and Campbell were loyal to Jobs and wanted to respect his privacy as much as possible. Most of the other members supported their discretion. "It was really up to Steve to go beyond what the law requires, but he was adamant that he didn't want his privacy invaded. His wishes should be respected," Gore would later say. "We hired outside counsel to do a review of what the law required and what the best practices were, and we handled it all by the book. I sound defensive, but the criticism really pissed me off."

At least one member had misgivings. As the head of the audit committee for a decade, Jerry York took corporate governance seriously. York was so disgusted by Apple's handling of the situation that he considered resigning. He ultimately decided against it because he knew it would cause a ruckus. He shared his concerns with almost no one. When his thinking later became public, the board found it hard to believe because York had never voiced his disapproval in a meeting. Some of his colleagues chalked it up to Monday morning quarterbacking. Officially, the board appeared to be unified.

Apple took unremitting heat for its veil of secrecy, but the commotion would have likely been worse if it hadn't coincided with a period of unprecedented growth.

Since the summer of 2008, the company had hit its stride in the mobile phone business with the launch of the iPhone 3G and the App Store. Its strategy to lure developers had worked beyond its wildest expectations, boosted by a revenue split model that allowed the developers to keep 70 percent of any app sales. Within months, Apple had thousands of apps in its store. A $2.99 app called iBeer that was created by magicians allowed users to tilt their phone like a mug and pretend to drink beer. iLightr was nothing more than a virtual lighter. But the ability to interact with them via the iPhone's big touchscreen provided a completely novel experience that was refreshing to consumers. And it wasn't long before more groundbreaking apps came along. Internet radio company Pandora's free, advertising-supported app let users pick a song, album, or artist and build a radio station around it on the go. The ninety-nine-cent Ocarina app made a wind instrument out of the phone by turning the microphone into an airflow sensor. All of these offerings added to the allure of the iPhone while the device's faster 3G Internet connectivity and cheaper price made it irresistible. Apple sold nearly 6.9 million iPhone 3Gs in the first three months as users collectively downloaded nearly 200 million apps. The more people downloaded apps, the more money developers were making, which in turn attracted media attention and even more developers to the App Store, which made the iPhone even more enticing. Apple had created a virtuous circle.

The growth was so spectacular that Jobs had made a rare appearance in Apple's quarterly earnings conference call with analysts that previous October.

"If this isn't stunning, I don't know what is," he said, adding that the iPhone outpaced sales of Research In Motion's BlackBerry device. "RIM is a good company that makes good products so it is surprising that after only fifteen months in the market, we could outsell them in any quarter," he said with false modesty. Apple's accomplishment was all the more remarkable, given that the rest of the world was experiencing the worst financial crisis since the Great Depression. Housing prices had plummeted, banks were facing total collapse, and consumer confidence was at an all-time low. While other companies in the tech sector such as Intel were closing factories and cutting jobs, Apple kept growing.

If Jobs had to go away for a while, January 2009 was not a bad time to do so. Despite the uncertainty over his health, investors were more optimistic than they had ever been about Apple. Its PR machine also worked behind the curtains to promote the idea that Cook and the rest of the executive team could steer the company just fine without Jobs.

That gave Tim Cook a huge advantage when he became interim CEO.

Jobs and Cook were a study in contrasts. One was a creative genius from California, known for his mercurial personality, and the other was an operations wiz from Alabama, known for his hyperrationality. Whereas Jobs might tell you that your idea was "shit" if he didn't like something, Cook took a more Socratic approach, asking question after question in increasing detail until he uncovered the weakness in an idea. Jobs had a presence that demanded attention. Cook's aura was more understated.

The difference between the two was precisely why Cook made an ideal deputy. Cook's expertise—from supply-chain management to customer support and inventory control—covered everything that his boss hated to do. As an operations expert, he set up a highly efficient system, so the beautiful prototypes that Jobs and Ive dreamed up could be mass-produced. Unlike some of the other executives who had passed through Apple, Cook also seemed content to stay behind the scenes, so he posed no threat to Jobs's star power. Unbeknownst to the public, Cook had been taking on more and more of the responsibilities of running Apple since he filled in for Jobs during his first medical leave of absence in mid-2004.

Aside from the occasional gossip in the media about Jobs's whereabouts, life without him at Apple was surprisingly normal. In the weeks after Cook took over the helm, he told an acquaintance, "I'm just going to stop listening to the press, turn it off, and get back to doing my work here at Apple." True to his word, Cook kept tight control over the company. Products were developed on schedule, and dealings with business partners continued as they had always done.

"We're just trying to do what we do every day," Schiller said in an interview that March.

Despite the cancer, Jobs continued to work on the company's most important strategies and products from home, and he regularly reviewed products and product plans, including the user interface for Apple's newest iPhone operating system. Cook and Ive met with him regularly at his home. But in general, executives were careful not to pull him into discussions unless it was absolutely necessary. Occasionally, someone would stand up and say, "Well, this is what Steve would do," as a tactic to get a point across, but Jobs was involved enough that not many people tried it. If he didn't like something, they still got emails or messages from him saying, "That's a dumb-ass idea."

One issue that Jobs concerned himself with that spring was the naming of Apple's new iPhone. Should it be iPhone 3G Speed? iPhone 3GS? Or iPhone 3G S with a space in between the *G* and *S*? Should the letters be capitalized? Italicized? Jobs ultimately approved the name iPhone 3G S, with a space in between the two letters, though he later got rid of the space.

Still, for all of his involvement, Jobs's absence from the public eye was conspicuous, particularly at occasions like the annual shareholders' meeting. Apple maintained its silence on his health throughout the spring, and its media relations department actively tried to discourage reporters from drawing attention to his absence. When the *Wall Street Journal* wrote a story in April about what Apple was like without Jobs, a spokesman demanded, "Why is this even a story?"

When Jobs went on medical leave, the tumors in his liver had already spread so much that he needed a transplant quickly. The problem was that the waiting list in California was one of the longest in the nation. The transplant network in the United States was closely monitored, and patients were ranked on a list using a complex algorithm that determined how critical the need was. But the law allowed him to sign up for a transplant in multiple locations as long as he could get there in time. Since he owned a private jet and distance wasn't an obstacle, he signed up in Tennessee, which had a much shorter transplant list than many other states. Jobs had a friend who knew the head of the transplant center at Methodist University Hospital in Memphis, ensuring that he would be well cared for.

Soon, renovations at a yellow, 5,800-square-foot mansion in Memphis's wealthy Morningside Park attracted the attention of curious residents. The area was a small neighborhood of about twenty homes with only one way in and one way out. Everyone knew everyone else. Many of them had been inside the five-bedroom, six-bath house before it was sold. They knew that the house had been well kept and updated. It had a large backyard with a beautiful pool. What kind of renovations could it possibly need? They realized the new owner was Jobs when a neighbor recognized his wife from an Internet search.

Jobs almost didn't make it, but on the weekend of March 21, 2009, a liver became available after a young man in his twenties died in a car crash. Jobs received the transplant and then stayed in Tennessee to recuperate until May, when he flew back with his wife and sister. They were met at the private airport terminal in San Jose by Cook and Ive, who joined them in a toast with sparkling apple cider. Jobs was eager to get back to work. He considered speaking at WWDC in early June, where Apple was launching the iPhone 3GS, but he decided against doing so because he still looked unhealthy.

When the *Wall Street Journal* found out about the transplant and called Apple's PR chief, Katie Cotton, she gave the standard response. "Steve continues to look forward to returning at the end of June, and there's nothing further to say," she said before quickly getting off the phone. Cotton and her team had frequently deployed every tactic in their playbook to challenge reporters. "You're going to discredit yourself and your publication," they would say. But this time, she behaved as if she was afraid to say anything misleading that would get the company into legal trouble later on.

Once the story went out, media outlets figured out where Jobs had been treated. Methodist University Hospital eventually confirmed the news.

I am pleased to confirm today, with the patient's permission, that Steve Jobs received a liver transplant at Methodist University Hospital Transplant Institute in partnership with the University of Tennessee Health Science Center in Memphis. Mr. Jobs underwent a complete transplant evaluation and was listed for transplantation for an approved indication in accordance

*with the Transplant Institute policies and United Network for
Organ Sharing (UNOS) policies. He received a liver transplant
because he was the patient with the highest MELD score (Model
for End-Stage Liver Disease) of his blood type and, therefore,
the sickest patient on the waiting list at the time a donor organ
became available. Mr. Jobs is now recovering well and has an
excellent prognosis.*

It was the perfect solution for Apple, which knew its disclosures
were being closely watched by corporate governance watchdogs. By
having the hospital confirm the details, Apple itself could continue to
avoid the obligation to provide details about its CEO's health.

In the aftermath, the company received much criticism about its
handling of the situation. Prominent experts like former SEC chairman Harvey Pitt questioned why Apple didn't disclose the transplant
sooner.

"We haven't gotten to the point where liver transplants are viewed
as routine surgery," he said in an interview, adding that he had questions about the board's processes.

Inquiries eventually died down. As unsatisfying as its disclosures
may have been, Apple's board and attorneys had done everything by
the book.

Jobs was motivated to get better quickly, and as part of his regimen,
he took many walks. One day, he invited the *Wall Street Journal*'s star
technology reviewer, Walt Mossberg, to accompany him.

Mossberg, who had known Jobs for decades, spoke to him often
and was one of the select few who received previews of new products.
It was sunny and warm when Mossberg pulled up to Jobs's house in a
quiet, unassuming Palo Alto neighborhood. The home looked like an
English cottage. It was large and attractive, but not unreasonably so.
A variety of apple trees, including the Macintosh, stood in the front
yard.

Mossberg and Jobs set out toward a park a few blocks away. As
they walked, they passed other homes. Many of them were relatively

modest in size, and none were ostentatious, but if you looked closely, there were signs of wealth—hidden security cameras, exotic trees on their lawns, luxury cars in their driveways, and a few homes with gates. But the neighborhood was friendly. Residents set out baskets of free apples from their trees in autumn and put up handmade signs that invited passersby to come into their garden and sit.

Jobs loved to gossip, so he and Mossberg talked about products and the industry. Sometimes he would make extreme assertions about people to get a reaction out of the journalist. Other times he would ask for an opinion. Jobs told Mossberg that he took walks every day and that each day he set a farther goal for himself. Today it was the neighborhood garden a half mile away. At one point, Jobs looked pale, and suddenly stopped. Mossberg, who didn't know CPR, could see a headline in his head: "Helpless Reporter Lets Steve Jobs Die on the Sidewalk."

As the pair arrived at the Elizabeth Gamble Garden, they were greeted by the delicate scent of the blooms from the sweet osmanthus tree. The talk became reflective. Sitting on one of the benches, they spoke about life, family, and illness. Mossberg had had a heart attack some years before, and he still suffered from diabetes. Jobs had always had a soft spot for people who were ill. Some years ago, when a teenager with muscular dystrophy visited Apple as part of his Make-A-Wish Foundation request, Jobs showed a gentleness that he rarely exhibited, taking the boy and his parents to lunch at the cafeteria, showing them around, and sitting with them in his office.

But now being ill himself, he was even more sensitive to health issues. Perhaps because of everything he had experienced, he didn't want others to make the same mistakes that he had made. Jobs lectured Mossberg about staying healthy.

Then they walked back.

3

Vertical

In January 2009, in the midst of uncertainty about Apple's future, the heir apparent to the throne asserted himself in a way he'd never before dared. Just a couple of weeks after Jobs started his medical leave, Tim Cook defined his own vision for Apple during a conference call with Wall Street analysts.

The call began like any other. Apple's chief financial officer, Peter Oppenheimer, ran through the company's quarterly performance. Then Cook joined him to take questions. Typically, the analysts focused their queries on the company's products, services, and strategies. This time the circumstances were far from business as usual. Jobs was on sick leave without much of an explanation, and rumors that he had cancer again were rampant. Few were surprised when the first question was about Cook's leadership.

"I just want to know how you'll run the company differently," asked Ben Reitzes, with Barclays Capital. "Tim, do you feel like you would be the likely candidate if the worst-case scenario were to happen, were Steve unable to return?"

Investors were already familiar with the broad strokes of Cook's background and expertise, and they knew he could operate the company without a problem. But that wasn't what Reitzes was asking. What he and the others really wanted to know was—did Cook have what it took to be Apple's CEO?

Oppenheimer jumped in immediately to give Apple's standard answer: "Steve is the CEO of Apple and plans to remain involved in major strategic decisions and Tim will be responsible for our day-to-day operations."

Cook could have left it at that, but for some reason on that day, the normally reserved executive was moved to step out of his mentor's shadow. "There is extraordinary breadth and depth and tenure among the Apple executive team," he began in his usual steady, quiet tone. "We believe that we are on the face of the earth to make great products and that's not changing. We are constantly focusing on innovating."

He then summarized Apple's principles. "We believe in the simple not the complex. We believe that we need to own and control the primary technologies behind the products that we make, and participate only in markets where we can make a significant contribution."

As he spoke, Apple's conference call service recorded every word.

"We believe in saying no to thousands of projects so that we can really focus on the few that are truly important and meaningful to us. We believe in deep collaboration and cross-pollination of our groups which allow us to innovate in a way that others cannot."

The conviction in his voice grew stronger as he emphasized Apple's drive for excellence. "We have the self-honesty to admit when we're wrong and the courage to change. And I think regardless of who is in what job those values are so embedded in this company that Apple will do extremely well." Cook finished by stressing how Apple was doing the best work in its history.

Until now, investors hadn't really known Cook. They knew that he worked like a machine and was extremely smart. They also liked him. Cook always treated them with respect and took their questions seriously. But they didn't know what made him tick. Now they did. There was no question that he *got* Apple. What drove him *was* Apple. The spontaneity of his speech made its eloquence even more impressive and profound. In one instant, he silenced any doubters about his commitment and passion for the company. He also proved himself to be a better self-marketer than he appeared to be. His manifesto became known as the "Cook Doctrine."

Jobs was irate when he heard about the earnings call. He had told Nocera a year earlier that he was making a concerted effort to give his executive team more exposure to show the breadth of the leadership. But it depressed him that Apple was doing so well without him.

During the first three months of Jobs's absence, Apple reported a 15 percent increase in quarterly profit even as other companies continued to suffer from the recession. In the following quarter, it sold more than six times as many iPhones as it did a year earlier. Apple's shares rose more than 80 percent to $142 since late January.

In reality, it was impossible for any manager to effect change in a company so quickly. Any accomplishment of Apple's was still the result of the work that had been done before Jobs left. But because the growth happened under Cook's leadership, he received most of the credit for the company's strong performance. Apple employees were also happier under Cook, who gave teams more freedom over strategy and product development. Everyone wanted their CEO back; he was their visionary leader, after all. But the calmer work environment was a nice respite. Cook didn't yell and scream like Jobs.

Five months into Jobs's leave of absence, Cook became one of the most sought after candidates in executive searches. Worried about Apple's ability to retain him, Wall Street began throwing out the possibility of Cook taking over permanently and Jobs becoming chairman. At minimum they thought he should be invited onto the board of directors.

"At this point," said Piper Jaffray's Gene Munster, "losing Tim Cook would be a bigger deal to investors than if Steve Jobs stepped aside."

Such praise was intolerable for Jobs, who considered himself indispensable. As much as he loathed the attention on his illness, part of him found it reaffirming that the world considered his well-being so crucial to the company's health. He had been deeply disappointed back in 2004 when the stock fell only 2 percent after he announced his surgery.

"That's it?" he had asked.

Jobs returned to Apple at the end of June just as he had said he would. On his first day, he threw a series of tantrums, ripping people apart and tearing up marketing plans. When Jobs heard about the press's sterling evaluation of Cook's performance, he hit the roof. Cook had done an excellent job, but the leadership and skill he showed in doing so was unsettling. He was also still sore about the "Cook Doctrine." Jobs chewed him out in a meeting with other executives.

"I'm the CEO!" Jobs yelled.

Cook was unfazed. You didn't work for Jobs as long as he had without developing immunity to such outbursts. The number-two man obligingly slipped back into the shadows, seemingly content to let Jobs regain control over the company.

Jobs's reentry was rough on everyone. After months of freedom, employees had to readjust to Jobs's intense scrutiny as he again reigned over product development meetings. Though some had wondered if he'd become more laid back as a result of his illness, the answer was no, at least where work was concerned. If anything, he pushed harder than ever. His primary focus upon his return: the iPad.

After redefining computers, digital music players, and cellular phones, this was his next revolution. If Apple could pull off the iPad, the company would cement its reputation as one of the greatest of all time. A successful new product would also pull the public's attention back to Jobs and complete his legend after beating cancer for the second time.

Initially, Jobs and his team had considered the device to be primarily for the education sector. One initial idea to build a buzz around the iPad was to give them away to key influencers like Nobel laureates. Jobs had always had an interest in education because he thought of it as an important way to empower individuals. He had such strong ideas about the subject that he had even once toyed with the idea of opening a school for the children of Apple employees on campus.

But as development of the iPad continued, it became apparent that it could be used much more broadly. Amazon's Kindle was becoming a must-have device. YouTube was becoming even more popular. There were also many uses for corporations and institutions in industries like logistics and medicine. Once the iPad launched, developers were sure to come up with clever apps for the device just as they had with the iPhone.

"We're discovering new uses every day," Ive enthused in the months before launch. As the sales team emphasized, there was no reason to give them away. The device would sell itself. The marketing team was so confident that it raised the starting price of the device to $499 from its original plan of $399.

Not everyone shared that excitement. When Jobs showed a prototype at a management retreat, some executives privately wondered who would buy it. Why would you need one in addition to a laptop and phone? Who wanted to carry more stuff?

Jobs unveiled the iPad to the public for the first time on January 27, 2010, at the Yerba Buena Center for the Arts, in San Francisco. In the audience was the surgeon who had performed his liver transplant as well as the surgeon who had operated on his pancreas in 2004. They sat next to his wife, his son, and his sister. As Bob Dylan's "Like a Rolling Stone" rippled through the audience, the lights dimmed. Jobs appeared from the left side of the stage dressed in his usual black turtleneck and jeans. After soaking in the standing ovation for a moment, a beaming Jobs began.

"We want to kick off 2010 by introducing a truly magical and revolutionary product today."

There was no mistaking the importance he placed on the product he was about to unveil. "Everybody uses a laptop and/or a smartphone. And the question has arisen lately. Is there room for a third category of device in the middle?"

Jobs talked about how such a device would have to be better than a laptop or a smartphone at performing key functions like Web browsing and email. Otherwise, it would be no better than a netbook, he said. "Netbooks aren't better at anything. They're slow, they have low-quality displays, and they run clunky old PC software," he declared. "They're just cheap laptops. And we don't think that they're a third category device."

As he prepared to reveal the iPad, his voice dropped to a hush. "But we think we've got something that is. And we'd like to show it to you today for the first time. And we call it the iPad."

In the slide behind him, the word "iPad" landed with a thud, kicking up some dust between an image of an iPhone and a MacBook.

Jobs spent the rest of the presentation showing off its features. To demonstrate how the iPad could be used casually in a living room, he sat down on a black leather Le Corbusier armchair behind him and picked up the iPad lying on the Eero Saarinen side table next to it. It was just like him to have designer furniture ready for a stage demo that was meant to show how approachable the iPad was. "Using this thing is remarkable. It's *so* much more intimate than a laptop and it's so much more capable than a smartphone with this gorgeous large display!"

He saved the news of the price until the end of the presentation. Making reference to pundits who had been predicting that Apple would sell the device at $999, Jobs announced that he would sell it at $499. On the slide behind him, the phrase "$499" crashed down on an empty screen, shattering the number "$999." The crowd broke out into lengthy applause.

The cover of the next issue of the *Economist* featured Jobs holding an iPad, wearing a robe and a halo.

Jobs was a master evangelist. About a week later, he took a trip to Manhattan to meet with the *New York Times*, the *Wall Street Journal*, and *Time*. The meetings would serve two purposes. One goal was to help change the tide of public opinion. The other was to get them excited about the iPad, so they would commit to developing news apps for it. Having good content would be crucial in selling the device.

Jobs first met with the top fifty executives of the New York Times Company over dinner at Pranna, a fancy Southeast Asian restaurant near Madison Square Park. Jobs had always had an affinity for the newspaper because he shared its politics, and he had already started conversations with the management about a partnership. After ordering a mango lassi and a vegan penne pasta—neither of which was on the menu—he demonstrated the iPad and talked about how it could serve the future of media.

The executives were not so easily influenced. To make the newspaper available on the iPad, Apple wanted a 30 percent cut of the revenue. Even harder to stomach, Apple wouldn't share subscribers' information unless they explicitly gave permission to do so. That was potentially debilitating for publishers. Not only would it give Apple enormous leverage in the industry; it also would make it impossible for media companies to mine their subscriber lists for names, email addresses, credit card numbers, and other valuable information that helped them attract advertisers and target new offers to readers.

"If you don't like it, don't use us," Jobs said in response to a circulation manager who tried to argue with him. "I'm not the one who got you in this jam. You're the ones who've spent the past five years giving away your paper online and not collecting anyone's credit card information."

At News Corporation the following morning, he found Rupert Murdoch, the owner of the *Wall Street Journal*, to be more receptive. Over a private breakfast in one of the executive dining rooms, Murdoch and his digital chief, Jon Miller, told Jobs that they wanted the *Journal* to be available on the iPad on day one. It was welcome news for Jobs. He didn't share Murdoch's politics, but he respected his entrepreneurship. They also discovered a few common interests, such as their love for sans serif fonts. When they finished talking, they walked together to a meeting room on the third floor, where a couple dozen of the *Journal*'s top editors and managers were waiting to interview Jobs. According to conditions that were set by Jobs's staff, the meeting was off the record, and the invitation list was kept secret. The room was also stocked with wheat bagels, cream cheese, hot chocolate, Smartwater, and Yogi licorice tea—all items that his staff had ordered. As usual, Jobs barely touched any of it.

The first question challenged a decision that Apple had made. Why was it not supporting Adobe Systems' Flash Player? Flash was the most popular technology for delivering video on the Web and was heavily used for animation and advertising, but Apple was instead supporting an emerging open-standard technology called HTML5, which lacked some of Flash's features.

Apple's anti-Flash policy wasn't new. The iPhone didn't support the technology, either. But Flash had become more relevant with the iPad because people expected to be able to watch online videos on the bigger device. "We think it's an old and clunky technology," Jobs told the editors, adding that the software was the biggest cause of crashes on Macintosh computers. "We don't spend a lot of energy on old technologies. We spend a lot of energy where the vectors are going."

Jobs's response was unsatisfying for editors. While they were excited about the iPad's potential to showcase their content in a new way, the device's lack of support for a technology they depended on complicated things.

Jobs passed out iPads, so the editors could try them out for themselves. When he looked like he was about to wrap up the session, one person asked the question that had been hanging in the air. "Steve, how are you feeling and when will you come back to work full time?"

"I'm already back full time," he said, repeating a line he had used before. "I'm vertical and loving every day of it."

Afterward, when Jobs stayed to talk to editors, no one stuck their hand out to shake his hand. They had been forewarned that Jobs did not want to make physical contact with anyone. He was on immunosuppressant drugs after his transplant, and he was worried about contracting diseases.

All in all, the *Wall Street Journal* meeting was successful. Over lunch, Murdoch talked about the iPad with some of his executives. "This is a really significant device. It's going to change the media world," he said, telling them that he expected them to build an app that was better than anyone else's on the day of the launch. Soon after, Apple arranged for a few iPads that the newspaper could use to develop its app. According to Apple's rules, the iPads were enclosed in a bolted case and chained to a table in a windowless room. A small team of developers worked so hard on the app that some of their colleagues joked that they were missing spring.

When Apple released the iPad in early April, analysts were disappointed by the initial sales. Apple had sold just three hundred thousand of them on the first day. The company considered the results to be respectable, but the figure was significantly lower than Wall Street's estimates. As with any brand-new product, it took time before people understood the iPad's use enough to want to make room for it in their lives. Some of Apple's most important products, including the iPod and iPhone, had had slow starts.

Unsurprisingly, many of the early adopters were in the technology industry. A lot of them were app developers in Silicon Valley, who found an immediate use for the iPads as a travel companion on the plane or a sleek tool for presentations when they were pitching venture capitalists. Die-hard Apple enthusiasts parroted, "It's *so* magical and revolutionary!"

Public opinion about the iPad began changing once reviewers tried out the device. Mossberg and the *New York Times*'s David Pogue both raved. Michael Arrington, the influential editor and founder of blog site TechCrunch, declared it better than his most optimistic expectations. "This is a new category of device. But it also will replace laptops for many people," he wrote. Though Jobs had demonstrated the iPad as a couch device, Arrington noted that it was "a perfectly usable business device."

Apple sold a million iPads in less than a month, twice as fast as the first iPhone's sales.

A bitter fight, however, was brewing between Apple and Adobe over Apple's lack of support of Flash. Jobs didn't even want developers to use Adobe software that would allow them to deliver Flash applications as native iPhone and iPad apps. Without Flash, users would see a blank space instead of the videos or interactive content they were supposed to see.

Publicly, Jobs argued that Apple wanted apps that were tailor-made for its devices. The company didn't want apps that compromised on features just so they would be compatible with all of the devices in the market. Jobs also claimed that Flash performed poorly on mobile devices, shortened their battery life, and created security problems.

His private reasoning was likely more complex. Allowing Adobe into the iPhone and iPad ecosystem would risk creating competition on two fronts. If developers could create software using Flash and deliver it directly to consumers over the Web, they could bypass the App Store—and its 30 percent tithe—entirely.

Also, Jobs probably hoped to lock developers into the Apple platform. Rivals were opening mobile app stores of their own, and he didn't want to make it easy for software programmers to create apps for them.

With the future of its business at stake, Adobe fanned the development community's discontent by calling attention to the issue on its corporate blog. Old grudges made the arguments particularly heated and personal. Jobs had never forgiven Adobe for not embracing the Macintosh when he was trying to turn Apple around in the late 1990s.

Adobe was "lazy," Jobs said at an internal Town Hall meeting shortly after he unveiled the iPad. "They have all this potential to do interesting things but they just refuse to do it."

Adobe supporters accused Apple of big-footing. Nearly eleven thousand Adobe fans banded together to support a Facebook page called "I'm with Adobe."

The battle came to a head when Jobs posted a sixteen-hundred-word essay called "Thoughts on Flash" on Apple's website in late April. It was a rare but not unprecedented move. The last time he had published something like that was three years earlier, when he urged the music industry to let Apple sell music without anti-copying protection.

"Flash was created during the PC era," Jobs wrote. "The mobile era is about low power devices, touch interfaces and open web standards—all areas where Flash falls short. . . . New open standards created in the mobile era, such as HTML5, will win on mobile devices (and PCs too). Perhaps Adobe should focus more on creating great HTML5 tools for the future, and less on criticizing Apple for leaving the past behind."

Apple ultimately triumphed. At the end of the day, the iPad was an Apple product. If you wanted to be a part of it, you had to abide by their terms. Flash had been on its way out anyway. Apple may have accelerated the transition, but the Web was already moving toward HTML5. The furor dissipated as developers and media companies came to terms with their reality and began planning their strategy.

But the victory cost Apple. The fight with Adobe enforced the perception that Apple was turning into an eight-hundred-pound gorilla. Despite Jobs's justifiable reasons to exclude Flash from the iPad, Apple came across as an oppressor. The controversy tarnished the empire's sterling brand image. More would soon follow.

Apple's App Store would be scrutinized next.

By spring 2010, nearly two years after launch, the store was astonishingly successful. It offered more than 185,000 games and other applications, most of which were free or ninety-nine cents. Venture capital firms were pumping money into app companies, and an ecosystem of advertising networks and other supporting businesses was emerging. Apple didn't disclose how much the company was making from its cut of sales, but Shaw Wu, a longtime Apple analyst, put estimates as high as $1 billion annually. Google, Apple's next-biggest App Store competitor, couldn't even come close with its thirty thousand applications, most of which were free. As with everything else the company did, Apple retained absolute control by requiring developers to submit their apps for approval.

The dominant power that Apple held in this new market was alarming for antitrust regulators. By this time, Apple had sold more than 85 million iPhones and iPod touches. The company's share in the overall mobile phone market was still small, but it was number three among smartphone makers and gaining. Apple's pop culture influence was even more ubiquitous. References to MacBooks, iPhones, iPod touches, and iPads appeared everywhere, overshadowing every other mobile phone maker.

Apple's convoluted app approval process added to the concerns. The company provided no clear guidelines about what they wanted to see, and apps were routinely rejected or ignored with little explanation. In a high-profile case, the company initially rejected Google's Voice app, which allowed users to make phone calls via the Internet. Apple approved the application only after Google complained to regulators about the anticompetitive move.

While Apple warred with Adobe, the two agencies responsible for enforcing federal antitrust laws, the Federal Trade Commission and the Justice Department, began holding discussions over which one of them would launch an inquiry into how Apple ran its App Store. At issue were two moves that Apple had made.

The first was the decision barring developers from using Adobe software that made Flash-based apps compatible with Apple's system. The second was a change in the developers agreement that banned apps from transmitting certain technical iPhone data to third parties. Apple had made the latter move in reaction to an analytics firm that abused the terms. But the action potentially prevented advertising networks from being able to target their ads, raising suspicions that Apple was giving its own mobile advertising service, iAd, an unfair edge.

In June, the Texas attorney general began making antitrust inquiries into Apple's relationship with book publishers over iBookstore, homing in on a creative pricing agreement between the parties.

Apple had always been an aggressive competitor and negotiator. But the kind of hardball tactics that had helped the company dig itself out of a hole and make a comeback were no longer acceptable now that Apple was top dog.

Meanwhile, Apple was soaring. In addition to putting the iPad on sale, the company was simultaneously working on the iPhone 4, which promised to wow consumers with a new design. Jobs lived for these unveilings, so he was shocked when his next hand was tipped in a bizarre turn of events.

In late March 2010, Gray Powell, a twenty-seven-year-old software engineer for Apple, went with his uncle to Gourmet Haus Staudt, a German beer garden in Redwood City, about a twenty-five-minute drive from campus. He had with him a prototype of an iPhone 4 that was disguised as the preceding 3GS model so he could field-test the device.

Powell was having a good time. Only a few years out of college, he was employed by one of the hottest companies in the world, working on one of its most important projects. He was smart and likable. "I underestimated how good German beer is," he enthused on his Facebook page, most likely on the iPhone 4 he was carrying.

But when he left the establishment later that evening, he left the prototype behind. It's unclear how this happened. Powell's last memory of the phone was inside his bag on the floor by his feet. But the phone was discovered on a bar stool, ending up in the hands of Brian Hogan, a twenty-one-year-old college student who had been enjoying a drink next to Powell. Hogan checked out the contents of the phone and waited for the owner to come back for a little while before he went home. Soon after, the phone was bricked, wiped remotely by Apple. As Hogan inspected the device, he realized for the first time that the iPhone was different from other 3GSs. When he took the case apart, he found a black phone with a flat back, a front-facing camera, and an aluminum frame. On the back was Apple's familiar logo with the word "iPhone" toward the bottom. In the space that would have usually shown the memory size in gigabytes, it just said "XXGB." It looked like a prototype.

Hogan knew that the phone belonged to Powell because he saw his Facebook profile on the phone. A quick search on the Internet showed that Powell was an engineer at Apple. Hogan made an attempt to contact Apple's customer care number, but when they wouldn't take him seriously, he decided to shop it to the media.

Staffers at the tech blog Gizmodo didn't notice Hogan's message at first because it had been sent to the site's general email address. When they finally spotted it, they were suspicious. Gizmodo was constantly receiving emails from people claiming to have information about unannounced products. Most of them were fake or outdated. But Jason Chen, Gizmodo's second-in-command, responded to the message just in case it was real. When Chen met with Hogan at a Starbucks in Palo Alto, he found a typical college kid who needed extra cash. As Chen examined the phone, Hogan chatted about working at a golfing range.

"Okay, this looks good from the surface, but I can't boot it up," Chen told Hogan. "It's hard to tell this is the real thing, so I need to take it apart and verify it." The two struck a deal for Chen to borrow the phone, take it apart, and report on it if it turned out to be real. Hogan wanted ten thousand dollars, but Chen negotiated him down by half.

About a week later, on April 19, Gizmodo published a piece called "This Is Apple's Next iPhone." The story received more than ten million hits. An hour after the story went live, Gizmodo's editor Brian Lam received a call from Jobs.

"Hi, this is Steve. I really want my phone back," he said. "We need the phone back because we can't let it fall into the wrong hands."

Lam played coy. First he denied having the phone. Chen had it. Then he refused to give the device back unless Apple officially claimed the prototype.

"Right now, we have nothing to lose. The thing is Apple PR has been cold to us lately. It affected my ability to do my job right at iPad launch. So we had to go outside and find our stories like this one, very aggressively," he wrote in an email to Jobs. Likening Gizmodo to the "old Apple," he pointed out that he was just trying to survive.

Up until this point, public sentiment was leaning toward Apple. Hogan shouldn't have sold a phone that wasn't his, and Gizmodo was coming under strong criticism for its checkbook journalism. Paying for news was something tabloids did. Not mainstream news media.

Jobs lost his patience. After complying with Lam's terms by sending a letter from his legal department and retrieving the phone from Chen, Apple pressed charges with the San Mateo police. On April 23 at around 9:45 p.m., Chen came home from having a hot-pot dinner with his wife to find his garage door ajar. When he tried to open it, officers came out and informed him that they had a search warrant. With his wife watching, they placed his hands behind his head and searched him for weapons. The cup of mango and lychee frozen yogurt they had been sharing melted in the car. Officers went through the entire house, taking eighteen items including computers, business cards, hard drives, digital cameras, and cell phones. They ignored Chen, who protested that he was a journalist and thus protected from such searches.

When Chen published his experience, there was an uproar. Even though Apple officials were not part of the raid, the company had triggered it. As unethical as some might have considered Gizmodo's decisions, a criminal search against a journalist crossed the line.

Jobs didn't care. When Apple officially unveiled the iPhone 4 at WWDC in early June, Gizmodo was excluded from the invitation list.

Though Jobs did his best to play up the iPhone 4 at launch, there was no getting around the fact that Gizmodo's report had stolen his thunder. The presentation was also marred by technical glitches as Web pages failed to load in demonstrations.

"Our networks in here are always unpredictable, so I have no idea what we're going to find. They are slow today," Jobs said, as the crowd went silent. A short while later, he asked everybody to get off the Wi-Fi network. Most of the audience assumed he was joking even though Jobs's tone was serious. By the time Jobs found out that personal Wi-Fi connections in the room were interfering with its cellular signals, he was seething.

"We figured out why my demo crashed," he said. "There are five hundred and seventy Wi-Fi base stations operating in this room. Okay? We can't deal with that."

The audience thought he was being funny again and laughed. Those who knew him heard his tone grow sharper and recognized his simmering anger. "We either turn off all the stuff and see the demos or we give up and I don't show you the demos. Would you like to see the demos or not?" he barked. "All you bloggers need to turn off your base stations, turn off your Wi-Fi. Every notebook, I'd like you to put them down on the floor. And all of you look around, I'd like you to police each other. . . . Come on! Look around you." Given that the media had been invited there to report the news of the day, it was a ludicrous request. But Apple staffers raced around the room, ordering everyone to close their laptops. They knew that they would hear about it if people didn't comply.

Later, when a demo video call with Jonathan Ive froze up, Jobs accused the audience of breaking their promise. When Ive asked him how he was doing, Jobs responded: "I'm doing okay except for these guys that aren't turning their Wi-Fi off."

Jobs was usually polished and professional during these product launches, but faced with an unforeseen predicament, the tyrant showed himself.

Fans around the world queued up to buy the phone. In San Francisco, the line started days in advance. One man toward the front of the line brought a tent for sleeping, a chair, books, and extra shirts. A woman farther down the line lugged a movie projector for after-dark entertainment. In Tokyo, customers, including a few in iPhone costumes, braved the sweltering summer heat. Apple Store employees passed out food and drinks.

But more trouble headed Apple's way. Customers immediately began complaining about the iPhone 4's poor cellular reception when the bottom-left corner was covered with their hand. Apple's first response was dismissive.

"Gripping any phone will result in some attenuation of its antenna performance with certain places being worse than others depending on the placement of the antennas," the company said, advising people to avoid covering the lower part of the device or get a case. Apple seemed to miss the irony of a design-centric company suggesting that customers cover up their phones.

As tech blogs ran tests confirming the issues, the cacophony grew louder. Lawsuits alleged that Apple had sold a defective product. Motorola ran advertisements for its upcoming Droid X, touting the phone's superior antenna design.

Apple could no longer deny the problem. In an open letter to iPhone 4 users, the company claimed to be surprised by the reports of the antenna issues and explained that a software problem was causing phones to mistakenly inflate readings of their cellular strength. While Apple didn't say so explicitly, it implied that the poor cellular quality was a network issue.

This was a partial truth. The problems tended to occur where cell signals were weakest, but the design of the phone itself contributed to the situation. Contrary to Apple's assertion, developers had known of this risk from the very beginning when Jobs and his design chief Jonathan Ive chose a stainless-steel frame that doubled as the antenna. The two had fallen in love with the design's simplicity, but it was a technical nightmare because metals interfered with signals, while the human touch degraded the quality of the cellular reception. The beauty of this kind of loop antenna had attracted designers in other companies, too, but there was a reason why no one had incorporated it in a product.

As the complaints mounted, *Consumer Reports* dropped a bombshell, saying that the publication couldn't recommend the iPhone 4.

Jobs was at his favorite vacation spot, Kona Village Resort, in Hawaii when he heard about the report.

"They want to shoot Apple down," he told board member Art Levinson. "Fuck this, it's not worth it," he said.

Cook worried that Apple might look complacent and arrogant. Though he normally deferred to Jobs, Cook persuaded him to address the issue. Pulling Jobs back from extreme positions was an important role of Jobs's executives, but few had the tact and sangfroid that Cook did. It was one of the reasons he was so effective.

Jobs flew back from Hawaii. Four days later, Apple held a press conference at its Town Hall auditorium. Still missing from the audience was Gizmodo.

Before Jobs came onstage, Apple played something called the "iPhone Antenna Song," which the staff had found on YouTube.

"There's an awful lotta hoopla around the iPhone antenna," the lyrics went. "If you don't want an iPhone 4, don't buy it. If you bought one and you don't like it, bring it back." The song set the tone for the press conference. When Jobs came out, he acknowledged the problems, but his tone was defiant.

"We're not perfect. We know that. You know that," he said. "But we want to make all of our users happy. And if you don't know that about Apple, you don't know Apple."

He then explained that Apple had been investigating the problem for the last twenty-two days. "It's not like Apple's had its head in the sand for three months on this, guys."

Jobs argued that the antenna problem wasn't unique to Apple. He showed videos demonstrating how it could happen with other phones by Research In Motion and Samsung. "This is life in the smartphone world. Phones aren't perfect," he said.

Though he offered free iPhone cases to appease unhappy customers, the gesture lacked sincerity. "This has been blown so out of proportion that it's incredible," he said. "There is no Antennagate."

When asked whether he would apologize to investors, his answer was no.

"You invest in the company we are, so if the stock goes down five dollars . . . I don't think I owe them an apology."

4

Attila the Hun of Inventory

That April, the comedian Jon Stewart directed his arrows at Apple. On the *Daily Show*, he addressed Apple's heavy-handed response to Gizmodo's possession of the iPhone 4 prototype.

"The cops had to bash in the guy's door?" Stewart asked incredulously after recapping the story in a video clip titled "iGoons." "Don't they know there's an app for that?" Behind him flashed a screenshot of a mock app, Ram iT.

"This whole thing is out of control," he declared as the graphic in the background changed to an Apple logo with the word "Appholes."

Stewart then called Jobs and Apple over to camera three, where he spoke to them directly.

"You know I love you guys, right? I love your products. I use them all the time. I even love your stores.

"Apple, you guys were the rebels, man, the underdogs. People believed in you. But now, are you becoming the Man? Remember back in 1984, you had those awesome ads about overthrowing Big Brother? Look in the mirror, man!"

Stewart held up an iPad so they could use its "surprisingly reflective" screen to see themselves.

"It wasn't supposed to be this way. Microsoft was supposed to be the evil one, but now you guys are busting down doors in Palo Alto while Commandant Gates is ridding the world of mosquitoes. What the fuck is going on?"

Stewart suggested that if Apple wanted to break down doors, it should start with its mobile service provider AT&T. "They make your amazing phone unusable as a phone. I mean, seriously! How do you drop four calls in a one-mile stretch of the West Side Highway?"

Or he proposed, "Why don't you kick in Paul McCartney's door for not letting us buy Beatles songs on iTunes?

"C'mon, Steve," Stewart implored. "Chill out with the paranoid corporate genius stuff."

Stewart's rant hit Apple where it hurt the most. The "1984" anti-establishment commercial that he had referred to still best represented Apple's core identity as an underdog. To accuse that company of turning into Big Brother, to suggest that Jobs was turning into the old Bill Gates, while Gates was off saving the world—Stewart couldn't have lobbed a bigger insult.

Apple was now an empire. Before the iPod, iPhone, and iPad, Apple's products had been primarily purchased by Apple fans, designers, and other discerning users willing to pay a big premium. Now Apple's products had become mainstream. Mac computers started at $999, the iPad at $499, and the iPhone 3GS at $199 with a subsidy. Apple was also selling the previous-generation iPhone 3G for $99, half of its original price.

By the end of June 2010, Apple was experiencing some of its strongest growth in recent history. Despite the initial skepticism, Apple sold 3.3 million iPads. Fears of the iPad cannibalizing computer sales were proved wrong as Apple sold 33 percent more computers in the quarter ended June than the year before. Concerns that the current-generation iPhone 3GS wouldn't sell in anticipation of the new iPhone 4 also proved unfounded. Phone sales increased 61 percent. Quarterly profit jumped 78 percent to $3.25 billion. Its stock was trading at around $250 per share.

Apple was morphing into the establishment. It controlled the market. That changed expectations. As an underdog, Jobs could stomp as much as he wanted. The bigger his target, the bolder he had seemed. But as a global giant, the aggression that had seemed puckish before now came across as dictatorial.

Apple's board recognized the need for an attitude adjustment.

"We need to make the transition to being a big company and dealing with the hubris issue," declared Art Levinson.

Jobs disagreed. "We are not arrogant," he insisted in board meetings. In his view, the start-up mentality he had instilled was essential to the company's success, and there was no way he would let outside forces dismantle it.

But realities were realities, especially when the government began taking a closer look at the company's practices. A few months later in September, Apple relaxed the restrictions on its App Store, so developers could use Adobe's Flash-based tool as well as other programming languages to write their apps. It also removed language from its developers agreement that prohibited mobile ad networks from collecting user data to target ads within iPhone apps. Responding to criticism about its opaque app review process, Apple published review guidelines for the first time. The remarkably candid document was at once humble and authoritarian, perhaps reflecting Apple's struggle with its changed status. It also sounded like Jobs himself had written it.

"We're really trying our best to create the best platform in the world for you to express your talents and make a living too. If it sounds like we're control freaks, well, maybe it's because we're so committed to our users and making sure they have a quality experience with our products," it said. "Just like almost all of you are too."

Some of the bullet points took on a sterner tone.

"We have over 250,000 apps in the App Store. We don't need any more Fart apps. If your app doesn't do something useful or provide some form of lasting entertainment, it may not be accepted," one said. Another: "If your App looks like it was cobbled together in a few days, or you're trying to get your first practice App into the store to impress your friends, please brace yourself for rejection. We have lots of serious developers who don't want their quality Apps to be surrounded by amateur hour."

The FTC's inquiry eventually dissipated.

Jobs still ruled over Apple. But Cook was the shadow warrior, his *kagemusha*. Cook was able to articulate Jobs's values so clearly in the "Cook Doctrine" because he had spent his career at Apple totally focused on Jobs's needs. Cook was the perfect steward to succeed Jobs when the inevitable happened. Unlike some of the others, who fancied themselves to be mini-S.J's, he was more likely to embrace Jobs's legacy as he guided the company toward continued prosperity.

Cook couldn't have been more different from Jobs. While Jobs exuded a titanic personality, Cook was impenetrable—quiet and self-contained. He started his career at IBM after graduating from college in 1982. Identified early on as someone with high potential, he rose quickly through the ranks. He worked his way up to director of North American Fulfillment in North Carolina, where he managed manufacturing and distribution for IBM's PC business in the United States, Canada, and Latin America.

"Tim was the first to work, the last to leave, and the smartest guy around the conference table," said Ray Mays, a former boss.

Cook stayed at IBM for twelve years, until he was offered an executive position at a small computer reseller in Colorado called Intelligent Electronics Inc. He rose quickly there as he helped restructure the company and nearly double its revenues. When the board accepted his recommendation to sell the company three years later, he was plucked by Compaq Computer Corporation, the industry's leading computer company and one of Intelligent Electronics' suppliers. Cook moved to Houston to run the group that purchased materials for products.

His colleagues at Compaq found him to be effective but aloof. While everyone else was in the suburbs, he lived in the city. While most of them had families, he was single.

Still, Cook liked Compaq and thought he could work there for a long time. Then one day a headhunter called. Apple was looking for a senior vice president of worldwide operations. "Why don't you come and meet Steve Jobs?" the recruiter suggested. Cook declined at first but was persuaded to at least meet him. He took the red-eye to California on a Friday evening and met with Apple's CEO the following morning. Cook later said it took him five minutes to change his mind and make the leap.

It was a surprising decision for someone reputed to be as rational as Cook was. In 1998, Apple's survival was far from certain. There was no logical advantage to leaving Compaq. A mentor warned him the move was foolish.

"Steve is a very compelling person. He has great vision," Cook said, shortly after accepting the job. "I saw this as a chance to participate in a corporate turnaround in one of the best companies in the world. The Apple brand is very powerful in the marketplace, especially in the design and publications industry, and in education as well. I saw this as a once-in-a-lifetime opportunity."

"My instincts have never let me down," he said. "This job felt right."

Apple needed new blood. In that spring of 1998, when Cook joined the executive team, the company was in the throes of restructuring, desperate for a capable executive who could work with Jobs and make Apple's manufacturing process more efficient.

Apple's operations unit was a mess. Long after its rivals had outsourced its manufacturing, the company continued to make most of its computers in its own factories, until demand plunged and the production costs became untenable. By the time Jobs returned to Apple, the company had sold off the bulk of its plants, but there were still hundreds of millions of dollars' worth of inventory lying around. The warehouses looked like Xanadu in the final scene of the movie *Citizen Kane*. Jobs wanted a lean operation, modeled after Dell. But better.

The trouble was that Jobs didn't have anyone to lead the team. He had gone through every qualified internal candidate in the first months after his return. Joe O'Sullivan, Apple's vice president of operations in Asia, was asleep one Saturday morning at his home in Singapore when the phone rang at six thirty. Jobs wanted him to come to the States.

"What about Jim McCluney?" O'Sullivan asked, naming the head of operations.

"McCluney's gone," Jobs said.

"What about Heidi?"

"She's gone."

"What about Vic?"

"He's gone."

"What about Sam?"

"I don't like him."

"What about John?"

"I don't like him, either."

"Ah," said O'Sullivan. "I'm the last man standing."

"No, no, someone over here said you'd be able to do it," Jobs protested. "I really want you to do it."

O'Sullivan had served as a vice president for only four months and knew he wasn't the right person. But his boss wasn't really asking. O'Sullivan got on a plane that weekend and reported for duty on Monday morning. He agreed to stay in Cupertino until Jobs found a more suitable replacement. O'Sullivan advised Jobs to hire someone with experience in procurement.

"We're going to be managing contract manufacturers," he said. "The whole business of operations is managing third parties. You need somebody who is very comfortable with that."

Jobs took his advice. Though Cook had been at Compaq for less than a year, his tenure there had coincided with a strategy shift that Apple wanted to make. Instead of manufacturing finished products ahead of time and letting them stack up on shelves, Compaq had switched to a build-to-order model, where it wouldn't start building computers until orders were received. This gave the company more flexibility and predictability in its manufacturing, but its success depended on the company's ability to manage suppliers and meet orders quickly and cost-effectively. Cook had been the man in the middle, who worked with Compaq's contract manufacturers to make that happen.

Cook was unflappable, making him the perfect counterbalance to the volatile Jobs. Executives who interviewed him for the job remembered how he coolly answered questions while munching on energy bars that he kept in his pocket.

Unlike his predecessors who sat with the operations team, Cook asked for a small office kitty-corner to Jobs's on the executive floor. It was a shrewd, chesslike strategy that allowed him to be four moves ahead. He had figured out that he needed to stay close to his boss in order to be attuned to his thinking. Few people thought much of it at the time, but they would later look back at it as an indication of the new leader's ambition.

From the beginning, Cook set colossal expectations. He wanted the best price, the best delivery, the best yield, the best of everything. "I want you to act like we are a twenty-billion-dollar company," he told the procurement team. It was a bold suggestion, considering that Apple only had about $6 billion in annual revenues and was barely eking out a profit. One person recognized the moment as a potentially significant turning point in Apple's history and carefully saved the notes from the meeting. They were playing in a new league now.

The first product the operations team shipped under Cook—the second-generation PowerBook G3—turned into a lesson on how he operated.

The black laptop computer, code-named "Wallstreet," was being assembled in Singapore, but it was running into serious problems ramping up production. Cook interrogated his staff in a manner that would soon become legendary. His questions grew more and more pointed until finally one of his deputies, Sabih Khan, offered to fly to Singapore to clean up the mess in person. A few moments later, when Khan spoke again, Cook looked at him.

"Why are you still here?"

Khan stood up and headed for the airport without a change of clothes.

The team soon got used to hearing "Who's on a plane?" and "Just fix it." Cook wasn't satisfied until someone could evaluate the situation with their own eyes.

Later when "Wallstreet" encountered another problem securing enough parts for production, Cook pushed the team to think creatively. "I want you to make them without the parts," he told one of his managers. It sounded like a joke, but the team soon found a way to build most of the computer without the missing piece, so they could cut down on lost time. The manager never again brought up an issue without considering every option.

Cook also tackled Apple's monstrous inventory. The company had already started whittling it down, but Cook considered any inventory to be fundamentally evil. "You kind of want to manage it like you're in the dairy business," he had said. "If it gets past its freshness date, you have a problem."

On his first trip to Asia, his operations team in Singapore thought it had thoroughly prepared for a meeting with him. At that time, Apple's manufacturing was managed regionally. The staff served Cook's favorite snacks—Mountain Dew soda and energy bars—and put together a presentation on inventory turns, a measure of how often a company sold and replaced its inventory. Asia was already turning inventory with an industry-leading standard of twenty-five times a year, up from a below-average performance of eight to ten turns previously. But Cook's reputation for thoroughness had preceded him, and the team was ready with a plan for how to get to an unprecedented one hundred turns. The higher the inventory turns, the more efficient the operation.

The meeting unfolded as predicted with Cook asking how they could get to one hundred turns. When they provided a satisfactory answer, he asked, "What about one thousand turns?" Assuming he was kidding, the team began laughing. He was asking them to figure out a way to turn inventory more than twice a day. Impossible.

"I'd like you to look at it," Cook told them.

Eighteen months after he started, Apple was turning inventory daily. Warehouses for components and raw materials were placed on the same factory grounds as the assembly lines, and the products were shipped directly to consumers whenever possible. Products were sent by air instead of sea to cut down on delivery time.

Within a few years the Asia team even accomplished near infinite turns, the epitome of inventory management. It meant they had no inventory. Its operations were so lean that customers were practically buying the product as it was coming off the assembly line. The company's overall inventory would rise again with the addition of new products and the opening of its own retail stores, but even then inventory levels stayed at significantly lower levels than its competitors' by a wide margin. Cook called himself the "Attila the Hun of inventory."

"It's not exciting to me to improve by five percent," Cook had said. "Now double or triple it, that's exciting."

Cook's way of dealing with suppliers was to be "aggressive and unreasonable." In his world, being reasonable meant compromise. Once when Motorola was having trouble delivering enough processors, Cook spent all day poring over spreadsheets, going through every step of the supply chain down to the sub-supplier. A colleague compared him to a moray eel. Once he clamped on, he didn't let go.

Cook was well liked and respected on the executive team, where his calm and rational approach was a welcome contrast to Jobs's tirades. But he was tough on those under him. Working for Cook required a

thick skin and a shared sense of perfectionism. Cook was never satisfied with the status quo, and he rarely doled out praise.

Some thrived in this unforgiving environment. Jeff Williams, the
head of worldwide procurement hired from IBM, could have been a
clone of Cook, down to his height and gray hair. A graduate of North
Carolina State University with an MBA from Duke, he was just as
much of an overachiever. When Williams first arrived, he worked so
hard that he slept in his office rather than go home. If there was a
problem in the supply chain, he flew to Asia to sort it out. Like Cook,
he was an intense perfectionist. He soon became a trusted deputy,
rising to vice president of operations. He was also frugal, driving the
same old Toyota Camry he owned when he first joined Apple.

Another manager who flourished was Tony Blevins, on the procurement team. A southerner like Williams and Cook, he was also a
thrifty IBM alum. Blevins didn't have the same commanding presence, but he embodied Cook's principles. He was so tough in negotiations that suppliers called him "the Blevinator." His basic stance was
that suppliers should be paying Apple for the privilege of its business.
When one of his subordinates declared an expensive gift certificate
from a supplier one year for Christmas, Blevins remarked on how they
clearly hadn't squeezed enough out of that company.

Others found the unrelenting environment intolerable. One of
them, Sheila Odle, was hired in early 2000 to manage product lines
for some of the computers. Odle, who came from a storage device company called Adaptec, was an experienced and competent planner. Yet
another southerner, she felt as if she understood Cook when she interviewed with him. But when she arrived at Apple, she had trouble
adjusting to the severe culture.

"How can you even think that that's okay?" Odle remembered
Cook saying when a team member reported on an issue. "You're an
idiot!"

Odle left Apple after a few months.

"I'm not from the school of berating and bullying to get what you
need," she said.

Cook's operations mercilessness was generating results. Four years
after he started working at Apple, he was given control of the sales

team. From there, he consolidated his power by gradually taking over noncreative roles like customer support and the Mac hardware business. In 2005, Jobs promoted him to chief operating officer.

The move surprised no one. By then, Cook had been working side by side with Jobs for seven years. No matter where he was traveling in the world, he never missed Jobs's Monday morning executive team meetings. He and the CEO were like yin and yang, interconnected and interdependent.

Cook's meticulous approach took a toll on those who worked under him. The team gave up evenings, holidays, weekends, and vacations to fulfill the promises that Cook made. When Jobs decided to change one of the colors in a new iPod shuffle lineup on a Christmas Eve six weeks before launch, Cook expected them to come in on Christmas Day to make it happen. In the early days, some of the old hands jokingly referred to themselves as "slaves."

When they weren't giving up their own time off, they were turning the screws on their Asian manufacturing partners to keep their factories churning around the clock. It was a particularly difficult task during Chinese New Year, the workers' most important holiday, when they expected to go home to their families.

Price negotiations with suppliers were emotionally grueling for all but the most hardened operations staff. People would sometimes break down in front of their Apple counterparts because they couldn't meet the company's pricing demands and were afraid of going back to their bosses empty-handed. One manager felt she was being deceptive when she extracted concessions from a supplier for a short-term contract, knowing that the other side was hoping for a permanent relationship.

Yet despite these efforts, the operations team rarely felt appreciated. They knew they had been relegated to the bottom of Apple's hierarchy, below the industrial design, hardware, and software units. Even before Jobs's return, operations had been described as a "bowel movement." Their lowly status only solidified under Jobs, who considered engineering, marketing, and design to be king.

During the holidays, the industrial design and engineering teams enjoyed lavish soirées at exclusive venues while the operations team was exiled to buffet lunches near Cupertino with iPods, iPads, and MacBooks as door prizes. When the first iPod touch launched, the engineering team was invited to the Exploratorium museum in San Francisco for a celebration with hors d'oeuvres, wine, and live music. Operations got nothing.

That gap extended into compensation. The rank and file in the operations unit received lower pay, bonuses, and stock allocations compared to their counterparts on other teams. Their salaries were respectable by industry standards, but as Apple became successful and their jobs grew more demanding, it didn't seem fair that designers could afford Aston Martins and multimillion-dollar homes in San Francisco while the serfs had to scrimp and save to buy a home in San Jose.

Amid such tensions, one manager found a crafty way to seek additional compensation. Paul Shin Devine, a global supply manager, was arrested in August 2010 for taking more than a million dollars in kickbacks from six Apple suppliers in South Korea, China, Taiwan, and Singapore. Devine was part of the iPod and accessories procurement operations team, responsible for selecting suppliers for parts and materials. Going back four years, he had provided suppliers with confidential information that would give them an edge in their negotiations with Apple.

To one Chinese company, he emailed step-by-step instructions on how it should negotiate with Apple to obtain money for new manufacturing tools that it needed to make Apple's products.

"I will propose 5 new sets," he wrote. "1) I will ask you to send me a quote for 5 new sets at full price ($120k x5=$600K). 2) I will then say too expensive. 3) I will suggest that you pay for 2 tool sets ($240K) and Apple pay for 3 tool sets ($360K). Which means I will try to get about $360K USD for Kaedar. What do you think?"

To another company, he recommended that it offer Apple a price of two U.S. cents for a part, in between the one cent that Apple was looking for and the four cents submitted by a competitor. Devine tried to avoid detection by using Hotmail and Gmail accounts and asking for payments in amounts less than ten thousand dollars, but he was eventually caught. The word among suppliers was that one of his contacts had felt exploited by the scheme and informed on him.

When the news became public, those who knew Devine were shocked. He was smart and well educated. Nothing about him had suggested any crookedness.

Devine, who pleaded guilty to fraud charges, was clearly on the wrong side of the law. But the case also provided a glimpse into the dark underbelly of Apple's success. Even though Devine made a good living—an average of $133,000 a year in salary and bonuses—he hadn't felt that his package was commensurate with the work he put into his job. Suppliers were desperate for any advantage in their negotiations with Apple. All of them were able to get away with it for as long as they did because Apple's massive scale of operations made it harder to keep a close track of its employees.

The dirty secret was that Devine wasn't the only one. Apple subsequently caught a handful of others who had done similar things and quietly showed them the door.

Apple's success brought pressure to the bottom end of the supply chain, too. In January 2010, a young factory worker in China jumped off his high-rise dormitory at a plant run by Foxconn, Apple's contract manufacturer. Ma Xiangqian had previously worked long overnight shifts, forging plastic and metal into electronics parts, but had been demoted to cleaning toilets after he broke some equipment by mistake. He had been earning the equivalent of one dollar an hour, working triple the legal overtime limit.

More than a dozen more deaths followed as other workers jumped off company buildings.

"Life is hard for us workers," said Ma Xiangqian's sister Ma Liqun shortly after her brother's death. "It's like they're training us to be machines."

The reasons behind the suicides were complex. Workers were certainly suffering as a result of the long hours and intense pressure they faced in their jobs. But a big part of it also had to do with China's social upheaval. Many of them were mere teenagers far from home and alone for the first time. It wasn't surprising that some of them would be emotionally unstable, especially if they were troubled by conflicts with coworkers, dorm-mates, or friends.

Those nuances, however, were lost as Foxconn's customers took a beating for profiting from the cheap, hard labor that these workers provided. Though many major technology companies did business with Foxconn, Apple received the most attention in the media because it was the highest-profile customer.

Unused to the limelight, Foxconn was slow and clumsy in its initial response. Then, as the scandal spun out of control, the manufacturer went on an unprecedented PR offensive and invited dozens of journalists to speak with its chief executive on the same factory grounds where most of the suicides took place.

"We're reviewing everything," Terry Gou told the reporters in a press conference as he apologized and bowed several times. "We will leave no stone unturned, and we'll make sure to find a way to reduce these suicide tendencies."

Afterward, Gou led a tour around the campus that included visits to dormitories, a hospital, a production line, and a counseling center. His aim was to show how well Foxconn treated its workers. But the visitors' eyes kept returning to the yellow mesh nets that Foxconn had begun installing around its buildings to catch jumpers.

It wasn't soon enough. Just hours later, a worker found a Foxconn building without a net and leapt to his death.

The Next Lily Pad

At dawn one brisk November morning in 2010, two empty buses rumbled in front of Apple's deserted corporate campus. As the drivers waited for their passengers, the chill gray of the parking lot was broken by the headlights of arriving cars. In Apple's intense culture, showing up early for work was not unusual, but the senior executives were gathering for a different mission. Instead of heading for their offices, they boarded the buses, idly chatting as they kept an eye out the windows to see who had been chosen to join them.

They were on their way to a Top 100 meeting, a secret corporate retreat hosted by Jobs at a resort south of Monterey Bay. Apple had just introduced a lineup of lighter and smaller MacBook Air laptop computers, and the company was heading into a busy holiday sales season. With new versions of the iPad and the iPhone also in the works, it was a good time to step away from the daily grind to think about Apple's future strategy.

Top 100s were meant for the company's brain trust. Everything about it was supposed to be kept hush-hush, and nobody was allowed to log it into their calendars. Those who made it onto the list were asked to keep the invitations to themselves, so their inclusion didn't spark jealousy. The secrecy added to the allure, reinforcing the sense that the company was working on things too exciting and special to share with just anyone.

In reality, the secrecy was a farce. There was no way that the disappearance of a hundred executives could go unnoticed, especially when they needed help from their staff to prepare. In their absence, some of the staff would hold a tongue-in-cheek "Bottom 100" get-together. It was usually low-key: lunch or a few drinks, snacks, and a bit of unwinding. One favorite place to go was BJ's Restaurant and

Brewhouse, which was so close that employees treated it like their private watering hole. They jokingly called it IL7, the unofficial seventh building on campus.

The core of the elite group included all of Jobs's inner circle of lieutenants such as Cook, Ive, mobile software head Scott Forstall, marketing chief Phil Schiller, and iTunes lead Eddy Cue. The rest of the handpicked names telegraphed Jobs's priorities and could change from year to year. Sales executives were largely excluded because Jobs viewed them as replaceable. Lee Clow, the creative lead at TBWA\Chiat\Day, the agency responsible for Apple's award-winning ads, was routinely invited even though he was an outsider. Jobs thought the hip and edgy campaigns that Clow's team dreamed up were crucial to Apple's brand. Intel executive Paul Otellini, as well as AT&T's key contact Glenn Lurie, had also attended a portion of the confabs in past years. The rumor was that Jobs preferred to mix it up by making sure that about a third of the list was composed of fresh faces.

Previous attendance was no guarantee of another invite. And even if you were chosen, your invitation could evaporate in an instant. One year, a new manager in the iTunes unit was pulled off one of the buses as it was leaving. After a meeting had gone poorly a few days earlier, Jobs had labeled him "an idiot" and ordered the hapless man disinvited.

Jobs called the Top 100 meetings irregularly and always with only about a month's notice. Some years there were two gatherings; other years none. Apple's biggest products and services were first unveiled internally at these meetings. Past participants heard about Apple's retail strategy and got early peeks at the iPhone and iPad. One year, Jobs solicited names for the digital music player Apple was developing. It was an exciting yet deflating moment.

After the audience eagerly suggested names like iPlay and iMusic, Jobs said, "Those are all shit. I'm sticking with what I've got."

A Top 100 invitation was a mixed blessing. To be chosen was unquestionably a privilege. But if you were asked to make a presentation, the pressure was nerve-racking. Jobs cared deeply about aesthetics, and those tastes extended to everything including slides, created on Apple's presentation software Keynote. Jobs had strict rules about them: one font family per presentation, three or five bullet points per slide, never four, and titles 30 percent above the center line. The file size also could not be more than eight megabytes, just enough to show up well on a projection screen. Jobs hated big files. Making the task all the more onerous, there was no instruction manual. Some had a knack for intuiting what their boss wanted. Bob Borchers, a product marketing executive for iPhones, had been so skillful that his colleagues used to call him the "Slide Bitch."

To get to the retreat, everyone had to take the chartered buses to the Carmel Valley Ranch, a five-hundred-acre resort surrounded by vineyards and lavender plants. It had a golf course, nine tennis courts, two pools, and two fitness centers. Not that these amenities mattered. The executives were there for work, not play.

This year, after the group arrived at the resort and checked into their rooms, they gathered in a large meeting room. Jobs kicked it off with a big speech.

"I'm going to give out a lot of secretive stuff here. I don't want to see a tweet or an email," he said. Jobs made sure there were no resort staff present. The executives knew enough not to bring computers into the room. Jobs threatened to fire anyone who broke his trust.

Over the next couple of days, each of Apple's business units took turns presenting to the entire group. The aim was to preview new products, explain strategy, and rally everyone behind what the company was doing. The sessions were intense, but there were no structured discussions, and attendees were free to ask questions or offer opinions. This created its own pressure. None of them wanted to embarrass themselves, but they needed to show Jobs that they were contributing.

In one of the first sessions, participants got to see the iPad 2 and the magnetic cover which latched on to the device and automatically woke the device when it was opened and put it to sleep when it closed. When it came time for the question-and-answer session, everyone wanted to know more about the cover, which could also fold into a stand to watch videos or make it easier to type.

"Enough with the fucking cover!" Jobs finally interjected. "Can we talk about the iPad?"

Another presentation addressed iAd, Apple's new mobile advertising service, which built on the company's $275 million acquisition of a mobile ad company called Quattro Wireless. Using one of his favorite analogies for describing essential products and services, Jobs introduced iAd as one of Apple's "tent poles." When the team's leader, Andy Miller, finished his presentation, one of the first questions came from Tim Cook, seated in the back.

"Andy, you know we bought your company and I know it wasn't the easiest integration," he said. "What can we do to make sure that we learn from this?"

"Well, the first thing I'd do is I'd recommend that the new guy doesn't report to this guy," Miller responded, pointing to Jobs.

For a moment there was dead silence. Then Jobs started laughing, and everyone joined in.

Cook's question was insightful because it wasn't about the product but about how to run the company. Historically, Apple had preferred to develop its own technologies. Acquisitions had been few, and there was little precedent for incorporating them into Apple. Newly absorbed staff often had trouble being accepted into the company's tight-knit culture. There was also no process in place to integrate them. Some showed up to find they had not even been assigned desks. As Apple acquired more companies, this lack of support needed to change. The question was one that would only occur to someone who was thinking deeply about the future.

The pace of the Top 100 was brutal. The days started with breakfast, and breaks were short. After dinner, many of the executives checked their emails and made phone calls even though they weren't supposed to be working during the retreat. Those who had presentations the following day would practice their lines over and over again.

The timing of the 2010 gathering was particularly heartbreaking for the baseball fans. The San Francisco Giants were back in the World Series for the first time since 2002, and the fifth game was scheduled that Monday, with the Giants ahead in the series 3-1. If they won this one, it would be their first World Series title since the team left New York in 1958. As beautiful as the resort was, Carmel Valley was the last place they wanted to be. As a small consolation, someone found a television and brought it into one of the halls. But they were called into dinner during the fifth inning. Neither team had scored yet. The fans tore themselves away from the action and reluctantly joined Jobs, who had been sitting alone in the restaurant, stewing. He had no interest in sports. When news of the Giants' victory spread through the dining room twenty minutes later, the fans made eye contact and rejoiced quietly so as not to further annoy their irritated CEO.

The Top 100s hadn't always been so serious. In the past, attendees had time to relax over drinks in the evenings. At some of the meetings, they played basketball. Schiller and Cue would often team up. One year, Apple's fearsome media spokeswoman Katie Cotton had supposedly dressed up as a cheerleader.

This time, the sessions were more somber. It was clear that Jobs wasn't feeling well. No one talked about it, but he was having trouble walking, and he looked depleted. One night, he left dinner early. The biggest sign that he wasn't himself was that he didn't chew anyone out. Past Top 100s could be dangerous encounters. Sometimes you couldn't even finish your presentation before Jobs interrupted. If he didn't like what you had to say, there would often be a feeding frenzy as other senior executives joined in the criticism. This time, the CEO was relatively quiet.

On the last day, employees were treated to a special moment. Jobs pulled up a chair in the front of the room.

"You've got Steve Jobs sitting right here," he said. "You're my guys. You can ask me anything you want. I don't care how dumb it is or insulting it is. I want to make you all feel comfortable about whatever questions you have about the company."

Hands shot right up. One of the most memorable questions was concerning rumors that had been spreading on the Internet about Apple's plans for a television.

"Are we going to make a TV?" someone asked.

Jobs didn't hesitate. "No," he replied. "TV is a terrible business. They don't turn over, and the margins suck." He added, however, that he wanted to own the living room.

Some in the audience took that as a sign that the Internet chatter was wrong, but those familiar with Jobs's cryptic ways weren't so sure. The veterans in the room interpreted his comments as a message to concentrate on what was before them now.

Jobs also talked about schools and how it pained him to see his kids walking to class with a heavy backpack full of outdated textbooks. He described the Apple TV digital media receiver as a hobby until the company got all the content. The question-and-answer session lasted about an hour and a half. Had the executives known that this would be his last Top 100, they probably would have kept him there longer.

Jobs knew that his cancer had returned. He was in pain again and having trouble eating. He was convinced that he was going to die soon and grew emotional at the thought that he might not celebrate any more of his children's birthdays. His son had just graduated from high school, and he had two younger daughters.

At Apple, he was involved in fewer projects. He paid attention to most things related to the iPhone or the iPad, but he was barely involved in the Mac business. One matter he insisted on controlling was the plan for a new corporate campus. Apple had purchased land that had previously belonged to Hewlett-Packard about a mile away from its current location, and Jobs had been working on the layout and design with the famed architect Norman Foster's firm. After many iterations, they had finally come up with a design that he thought expressed the company's values for generations to come.

One day in late 2010, Jobs called an impromptu meeting with the city's two top officials to float the idea of the new corporate headquarters. When Gilbert Wong, the mayor-elect, and Kris Wang, the outgoing mayor, arrived at Apple, they were led to the CEO's conference room. "Hi, my name is Steve," Jobs said as he greeted the starstruck council members, shaking their hands firmly. The officials hadn't known they were meeting with him until a few minutes prior. Jobs had probably orchestrated it that way, so he could catch them off guard and dazzle them into fast-tracking approvals for the new offices.

Recounting the days when he worked with the Emeryville City Council to build Pixar's campus, Jobs tried to persuade Wong and Wang to grant him an exception. He was even willing to pay an in-lieu fee. The two tried to show respect by addressing the CEO as "Mr. Jobs." But the answer they gave him was no. It was naïve of Jobs to even try. Democracy didn't work that way. In addition to requiring approval from three out of five council members, Apple would need to complete an environmental impact report, and the plans would need to be discussed in a public hearing. "He was trying to negotiate as if we were the decision makers," Wong recalled. "He was just selling, selling, selling, saying 'I want this,' 'I want that.'"

Jobs wouldn't give up. "I want to show you something," he said, suggesting they walk over to another room. When they got there, he made a show of taking out his key and inserting it into the locked door.

Inside stood a beautiful model of the new campus, with an abundance of windows and green, open spaces. One of its distinctive features, Jobs pointed out proudly, was that everyone would be treated equally and there would be no special CEO office. The city officials were impressed but stood firm. Jobs had to follow the process like everyone else. If it struck Wong as ironic that Jobs was touting his egalitarian layout to seek special dispensation from the city, he didn't mention it.

Jobs didn't get what he wanted that day, but he did succeed in making it an unforgettable visit for Wong. "I had an opportunity to meet with President Obama through the U.S. Mayors Conference and I got to personally meet with the president of Taiwan as well as U.S. senators, congressmen, and other government officials. I've been to the People's Republic of China and India, and I've met other corporate people in Silicon Valley, but the highlight would have to be my meeting Mr. Jobs," he said later. "How often does someone from the public have a one-on-one conversation with a CEO about a two-point-eight-million square-foot campus?"

Wong may have been awed by Jobs's presence, but he still had the wherewithal to notice how sick Jobs had looked. By then, Jobs was working more from home again. The official company line was that he was fine now that he had a new liver. Employees, who saw him in the cafeteria, knew better. By Christmas, Jobs was down to 115 pounds, fifty pounds below his normal weight. Every inch of his body, he told friends, felt like it had been punched.

In January, doctors detected new tumors. Jobs didn't want to go on medical leave but had no choice. On January 17, 2011, he sent out an email to all employees:

> *Team,*
>
> *At my request, the board of directors has granted me a medical leave of absence so I can focus on my health. I will continue as CEO and be involved in major strategic decisions for the company.*
>
> *I have asked Tim Cook to be responsible for all of Apple's day to day operations. I have great confidence that Tim and the rest of the executive management team will do a terrific job executing the exciting plans we have in place for 2011.*
>
> *I love Apple so much and hope to be back as soon as I can. In the meantime, my family and I would deeply appreciate respect for our privacy.*
>
> *Steve*

As before, the note was scarce on details. But it was different from the last one because Jobs didn't give a time frame for his return. The carefully selected wording implied that there was a chance he might not be back at all.

In retrospect, there had been warning signs. When Apple announced a new partnership with Verizon Wireless the week before, it was Cook, not Jobs, who showed up at the press event. A News Corporation press event where Jobs was slated to appear had been abruptly postponed. Nevertheless, the announcement of another medical leave took many people by surprise. The fact that he sent the notice on Martin Luther King, Jr. Day led some to speculate about whether the company had deliberately picked a day when the U.S. stock markets were closed. Apple shares fell 8 percent in Europe.

In an earnings conference call the day after the email, Wall Street refrained out of courtesy from asking about Jobs's health, but Cook seemed to detect the anxiety. When an analyst asked about Apple's long-term business planning, he assured him that Apple was doing its best work ever.

"We are all very happy with the product pipeline," he said. "The team here has an unparalleled breadth and depth of talent and culture of innovation that Steve has driven in the company. Excellence has become a habit, and so we feel very, very confident about the future of the company."

Eventually, the concern ebbed. It helped that Apple had reported a 78 percent surge in profit and record sales during the previous quarter. For the moment, Apple still revolved around Jobs. Even though he had formally removed himself from the responsibilities of running the company and wasn't as available as he used to be, his deputies, including Cook, were reluctant to make decisions without his approval. It was especially true on matters that they knew he would have wanted to be involved in. Halving the minimum spending for iAd campaigns from $1 million to $500,000 was one of them.

The plan had been in the works for weeks, but no one wanted to make a decision on his behalf. Everyone held their breath, hoping he would return.

That February, Jobs was preoccupied with the passage of time. His twentieth wedding anniversary was coming up the following month. He hadn't always paid attention to occasions like this in the past, but this year was different. Deciding to enlist some help, he called Tom Suiter, an old friend and a veteran designer.

"You remember how you helped me out with our wedding invitation?" he asked. "I want to do something really special for Laurene and the kids." When Suiter asked what he had in mind, Jobs responded, "I don't know yet. That's what I wanted to talk to you about. Something that could provide them with wonderful memories."

Suiter had worked for Jobs as a creative director in the early 1980s and had been part of the team that launched the "1984" ad campaign. He helped name NeXT, the company Jobs had founded after he was booted out of Apple in 1985. Jobs had originally wanted to call it "Two" because it was his second company. Suiter convinced him that it was a bad idea because everyone would want to know what happened to "One." After cofounding the ad agency CKS Partners, Suiter continued to work for Jobs. He had been involved in the famous "Think Different" ad campaign and the initial designs for the Apple Stores. Even after they stopped working together, the two stayed in touch.

On the day of their first meeting, Suiter arrived at Jobs's house with some watercolor paper and a set of Conté crayons. Suiter's idea was a handmade white linen box that would open up to a black linen container with twenty photographs. He envisioned that the box would have some kind of logo on it. Suiter was hoping to convince Jobs to design that himself. When he suggested it, Jobs refused to consider it.

"Steve, you could hire anybody on the face of the planet to do this for you. I'm so appreciative that you asked me to do it, but I just think it would be so cool if you could do something," ventured Suiter as they sat in Jobs's atrium. "What if it was just this really nice little heart with a two and a zero on it?"

For a moment Jobs was silent.

"I'll give it a shot," he said slowly. "But will you draw it first and I'll copy it?"

After Suiter did a few hearts, Jobs picked up a crayon. He very carefully drew one side and then the other. And then he drew a 2. The crayons provided beautiful texture, but Jobs didn't like what he saw and tried to discard them.

"Don't worry about it. Do a few and then I'll be able to Photoshop it and put them together," Suiter told him.

As they got together over the next several weeks, they talked about life and compared notes about their children. Sometimes, the conversation was lighthearted. They would talk about funny moments or their kids' Halloween costumes. Suiter used to make Superman and werewolf costumes for his sons.

Jobs expressed regret that he hadn't been a better father. He wished he better understood his daughters.

One time as Suiter was walking out, Jobs told him goodbye. The formality of the farewell frightened Suiter so much that he began praying as soon as he reached his car. "Come on," he told himself. "No, this can't be the end."

He was immensely relieved when he saw Jobs again two weeks later. The two never talked directly about Jobs's legacy. Suiter didn't want to consider a future without his friend, and it was clear Jobs intended to be around for a while.

"Is it a month? Is it a year?" Jobs would speculate. "I don't know. It could be ten years."

Suiter caught him in a reflective mood once.

"Steve, I think about the life I led and I'm so happy," Suiter opened by saying. Like Jobs, Suiter had struggled with cancer a few years back. "You must feel the same way because you've lived a life of twenty men."

"I know," Jobs said. "I know. I have."

Suiter asked him about the coolest thing he'd done in his life. Jobs admitted that he hadn't thought about that kind of thing until recently.

"You know, I sort of feel like we don't step back and think about those kinds of things. Jony and I were talking about that," Jobs said, referring to Apple's designer Jonathan Ive, with whom he spent a lot of time. "For me, it would be a tie. Without a doubt the first Macintosh was so much fun to do, and I truly believe the first-generation iPhone was so similar in that it was just so different, so unique, and so far beyond what anybody else was expecting."

When Jobs finished designing the logo of the heart, Suiter found someone who made exquisitely crafted linen boxes. Then he located a museum quality printer. Suiter also persuaded Jobs to write a letter to his family that was letter-pressed and enclosed under a sheet of vellum paper. As before, Jobs was initially resistant.

"Aw man. Aw God."

"No, come on! You can write that," Suiter said to coax him. "Think about it. The first time you saw her she swept you off your feet, right?"

"Yeah."

The result was a sweet and melancholy note. When the gift was finally finished, Jobs had tears in his eyes.

"It's perfect."

Jobs kept the details of his health out of the public eye again. But this time, the world was better informed, based on what they knew of his medical history: If he was ill again so soon after his liver transplant, it was highly likely that the cancer had come back. This barest of information was enough cause for concern when his sickly image appeared online.

When Apple sent out invitations to the media for an event on March 2, 2011, it displayed a partial image of an iPad, hinting that the event would concern a new model. Fans were eager to know what it would look like, but the bigger question was, would Jobs be there?

Reporters braved the cold San Francisco rain to arrive early that morning at the Yerba Buena Center. When the doors to the auditorium opened, the media rushed in. The scanning to find Jobs began immediately. Reporters identified Cook, Schiller, Ive, and other executives standing near the front. If they were already in the room, who was going to be the master of ceremonies?

Anticipation mounted when the lights dimmed and George Harrison's voice purred from the loudspeakers, singing "Here Comes the Sun." Then Jobs walked out.

"We've been working on this product for a while and I just didn't want to miss today," he said somewhat bashfully to cheers and applause.

As word got around that Jobs did not have much time left, dignitaries made a pilgrimage to see him one last time. Google's cofounder Larry Page dropped by for advice on how to be a good CEO. Bill Clinton visited and talked about American politics and the Middle East. Bill Gates spent more than three hours with him in May, during which they reminisced and discussed the future of education. Sony CEO Howard Stringer called him with information about a possible drug treatment that he thought might be helpful.

Jobs had finally accepted the seriousness of his diagnosis. He would tell people that he was just trying to manage his cancer. He had been one of the first twenty people in the world to have all of his genes sequenced, so doctors could formulate specific drugs that could provide targeted therapy.

"These drugs work great, and then they stop," he explained to one friend. "Then it takes a while to find another drug."

Jobs had good and bad days. For his friends, the swings in his health were frightening. Ive was openly distraught, lingering as much as possible at Jobs's house. He fretted over his illness so much that dealing with him left Jobs exhausted. Cook visited often, but he was stoic, sticking to business, betraying little emotion. iTunes head Eddy Cue would go see the Boss—he was the only one who referred to Jobs that way—and come back shaken up. Jobs had picked him out when he was nobody and had entrusted him with enormous responsibility. Cue genuinely loved the man and didn't try to hide it.

"I just came back from seeing the Boss," he said after one visit. "He's not good."

"But he came back from this, he came back from this, he came back from this," he repeated, as if he were trying to convince himself. It sounded as though he'd had this conversation a number of times in his head. "The guy's a cat. He's got nine lives. We've got a few more to go. We can never count out the Boss."

When Apple held its WWDC conference in early June, it announced ahead of time that Jobs would attend. Apple fans entered the auditorium to the carefully vetted song of choice—James Brown's "I Got You (I Feel Good)." For an instant, some audience members became hopeful. Would they see a healthier Jobs? As the music died down, he finally came out onstage to a lengthy standing ovation.

"I love you!" somebody called out.

"Thank you," he responded as he smiled. He raised his eyebrows and pointed at those he recognized in the audience. "It always helps and I appreciate it very much."

He looked delighted to be there, but his presence onstage was noticeably short. He also appeared frightfully thin. The audience didn't even know that Jobs was wearing two shirts and thermal underwear.

The next evening, Jobs made an appearance at the Cupertino City Council. He had been scheduled to go to a U2 concert in Oakland but decided instead to unveil his plans for the new corporate campus. Failing to circumvent the government process the previous winter, Jobs wanted to announce the project formally before the public. He had already canceled plans to attend a council meeting once earlier that spring, and he didn't want to put it off again. He needed to be the one to sell the idea. The new office celebrated his success. Even its location—the former grounds of Hewitt-Packard, a company he had admired when he was young—was symbolic.

"Apple's grown like a weed," Jobs began, explaining how the company's current headquarters only held about a quarter of the twelve thousand employees it had in the area. Jobs then wove a spell as he talked about Apple's roots in Cupertino and what the area meant to him. He showed a rendering of the new campus with an enormous round building.

"It's a little like a spaceship landed, but there it is. And it's got this gorgeous courtyard in the middle, but a lot more," he said, boasting about the curve of the building. "There's not a straight piece of glass on this building."

Jobs wanted to make 80 percent of the property landscaping. A big parking lot would be installed underground. He would almost double the number of trees. "We've hired one of the senior arborists from Stanford actually who's very good with indigenous trees around this area," he told them. In a nod to Cupertino's beginnings as a place filled with vineyards and orchards, he wanted to plant fruit trees with apricots, apples, and plums. There would also be a three-thousand-person café, an auditorium, fitness center, and R&D buildings. Jobs painted a picture of the campus as an environmentally friendly facility, where they would generate their own power with the grid as a backup.

"I think the overall feeling is going to be a zillion times better than it is now with all the asphalt," Jobs said. "We do have a shot of building the best office building in the world. I really do think architecture students will come here to see this. I think it could be that good."

It was a dazzling performance that made one almost forget he was dying. There was one tense moment when Kris Wang, the former mayor who was now a council member, asked Jobs if Apple would give the city free Wi-Fi connectivity in return for approving his plans. The response was vintage Jobs.

"See, I'm a simpleton. I've always had this view that we pay taxes and the city should do those things," he said, condescendingly polite. "If we can get out of paying taxes, I'd be glad to put up a Wi-Fi network."

He reminded the council that Apple was the largest taxpayer in Cupertino. "If we can't build it," Jobs said referring to the new campus, "we have to go somewhere like Mountain View, and we'd take our current people with us and we give up and over the years sell the land here."

Gilbert Wong was so excited by Jobs's presence that he had trouble forming a coherent sentence. "Thank you, Mr. Jobs. We're really excited that you call Apple our home," the mayor said at the end of his presentation. He had probably meant to say he was glad Apple called Cupertino their home.

Judging from the reaction that day, there was little doubt that Cupertino would approve the plan. But two questions hung in the air. Would Jobs live to see the new campus? And was Apple's new campus meant to be a monument to Jobs, his high-tech Taj Mahal? As Jobs described it in private conversations, the new building would be so big that the middle of it could hold St. Paul's Basilica.

By July, Jobs's cancer had spread to his bones and other parts of his body. He was in greater pain, and he stopped going in to the office. Though he was once sighted with a plate of French toast at a restaurant near his home, he was generally not sleeping well and could barely eat solid foods. He moved his bedroom downstairs. Whispers were starting to circulate around Silicon Valley. He had promised the public that he would step down when he was no longer fit to run the company. That time had come.

Jobs called Cook one weekend.

"I'd like to talk to you," Jobs said.

"Fine," said Cook. "When?"

"Now."

Cook drove over to Jobs's house. It only took a few minutes because they lived less than two miles apart. When Cook arrived, Jobs told him that he wanted his steward to be the next CEO.

"There has never been a professional transition at the CEO level in Apple," he said. "Our company has done a lot of great things, but has never done this one." Jobs wanted to set an example of how to transfer power right.

"I have decided, and I am recommending to the board, that you be the CEO, and I'm going to be the chairman."

Cook looked at his mentor. "Are you sure?"

"Yes," said Jobs. When Cook asked again, Jobs told him not to ask him anymore. His mind was made up.

The two then spoke about what that would mean. Jobs was very clear about what he didn't want.

"I saw what happened when Walt Disney passed away. People looked around and they kept asking what Walt would have done. The business was paralyzed and people just sat around in meetings and talked about what Walt would have done."

In a refrain that would later be repeated throughout the company, Jobs then advised Cook, "I never want you to ask what I would have done. Just do what's right."

The date Jobs chose to make the announcement was August 24, 2011, the day of the company's board meeting. Jobs wasn't strong enough to get to the office himself, so arrangements were made to have him driven there and taken into the boardroom as discreetly as possible by wheelchair.

Jobs arrived just before 11 a.m. After the board members had finished with their committee reports and other routine business, Jobs asked for a moment because he had something personal to say. They knew what was coming. Cook and some of the other executives who were present left the room. This was a board matter.

He began to read from a letter he had prepared.

I have always said if there ever came a day when I could no longer meet my duties and expectations as Apple's CEO, I would be the first to let you know. Unfortunately, that day has come. I hereby resign as CEO of Apple. I would like to serve, if the Board sees fit, as Chairman of the Board, director and Apple employee. As far as my successor goes, I strongly recommend that we execute our succession plan and name Tim Cook as CEO of Apple. I believe Apple's brightest and most innovative days are ahead of it. And I look forward to watching and contributing to its success in a new role. I have made some of the best friends of my life at Apple, and I thank you for all the many years of being able to work alongside you.

A long silence descended. His friend Bill Campbell sat with tears in his eyes. Al Gore and Mickey Drexler acknowledged his contributions to Apple. Art Levinson praised Jobs for the smooth execution of the leadership transition. Some Apple shareholders had never believed that Apple had a succession plan because it had refused to disclose it. But there had been a plan in place since the day Jobs had been diagnosed in 2003.

Jobs stayed for lunch, during which Scott Forstall and Phil Schiller showed some mockups of products the company was working on. Jobs asked questions and shared his thoughts about features in future phones. When Forstall demonstrated Siri, a new virtual personal assistant app that Apple was planning to launch that fall, Jobs wanted to try it out for himself.

"Give me the phone," he commanded.

Jobs started with an easy question about the weather. Then he threw a curveball.

"Are you a man or a woman?"

The room broke into laughter as Siri answered.

"They did not assign me a gender."

Back home, Jobs found his wife harvesting honey from the hives they kept in the backyard with help from their youngest daughter, Eve. When they walked into the house with the honey pot, they were joined by his son Reed and other daughter, Erin. They gathered in the kitchen to taste their harvest. Jobs took a spoonful, pronouncing it "wonderfully sweet."

Apple had been his life for more than two decades. It was his company, yet it was no longer his to run. He still had ideas and there was so much he wanted to do. He intended to stay involved as long as he could, but the media was writing about his accomplishments as if he were finished.

Jobs picked up the phone to call the *Wall Street Journal*'s Walt Mossberg, who had been furiously pounding away on a story about his legacy. Jobs wanted to talk to someone about his mixed feelings and explain how he was stepping down but was still planning to stay involved.

After discussing his health, Jobs talked about Cook and the succession plan. He believed that he had a deep bench of executives on the leadership team and he was confident about Apple's future.

"You know, people criticize me for never publicizing it, but we did have one and that plan calls for Tim to become the CEO," he said. "I just thought rather than waiting until I died and having it done in chaos . . . I meant what I said in that letter. I don't feel I can do the job as fully as I need to. I have to narrow my focus. I have to be selective on certain things. There's a lot that I'm not going to be able to do."

But he wanted Mossberg to understand that he was still going to be working on projects. He alluded to products Apple was working on.

"I want you in a couple of months to come out," he told Mossberg. "I'd love for you to come out to see it."

As usual, the conversation was off the record, but when Mossberg hung up, he changed one thing in the piece he was working on.

"To be very clear, Jobs, while seriously ill, is very much alive. Extremely well-informed sources at Apple say he intends to remain involved in developing major future products and strategy and intends to be an active chairman of the board, even while new CEO Tim Cook runs the company day to day," he wrote. "This is not an obituary."

Jobs continued to feel somewhat melancholy about the transition. When his biographer Walter Isaacson asked him what it was like to leave the company he founded, he flew off the handle.

"Why do you think I'm leaving?" he demanded. "I just said in the statement, I just told you! I'm going to stay. I'm going to be advisor."

Even though he was very weak by then, he told Isaacson, "I'm going to be okay. I'm going to be okay. I'm going to get to the next lily pad."

Once Jobs announced his resignation, the secret was out. Jobs's time was limited. More calls and emails from friends, colleagues, and other people in the industry who knew him came pouring in. Jobs responded to a few of them, but he didn't seem to be checking his email very much anymore. He declined invitations to farewell dinners and ceremonies to accept various awards. His wife turned away most requests for a final visit. He saw only his closest friends. He watched a movie with Campbell. He had sushi at Jin Sho in Palo Alto with physician Dean Ornish, a well-known advocate of changing diet and lifestyle to prevent heart disease.

Jobs also offered advice to Apple executives on the unveiling of the iPhone 4S, scheduled for Tuesday, October 4. While Cook welcomed reporters to his first product launch event as CEO, Jobs lay dying. He called his sister Mona Simpson on the phone and asked her to come to Palo Alto. When he started saying his farewells, Simpson cut him off.

"Wait. I'm coming. I'm in a taxi to the airport. I'll be there."

"I'm telling you now," Jobs told her gently, "because I'm afraid you won't make it on time, honey."

When Simpson arrived, her brother was still lucid, surrounded by his wife and his children. Jobs told her that he was sorry they wouldn't be able to grow old together. Later that day, his breathing slowed and then became labored. That afternoon, he looked at his sisters, then his children, then his wife. Simpson later wrote that Jobs's final words were "OH WOW. OH WOW. OH WOW."

The death certificate cited respiratory arrest and metastatic pancreas neuroendocrine tumor as the cause. The news was announced in an email from Cook to Apple's employees:

Team,

I have some very sad news to share with all of you. Steve passed away earlier today. Apple has lost a visionary and creative genius, and the world has lost an amazing human being. Those of us who have been fortunate enough to know and work with Steve have lost a dear friend and an inspiring mentor. Steve leaves behind a company that only he could have built, and his spirit will forever be the foundation of Apple. . . . No words can adequately express our sadness at Steve's death or our gratitude for the opportunity to work with him. We will honor his memory by dedicating ourselves to continuing the work he loved so much.

Tim

Everyone had known Jobs was gravely ill. A photo that was published on celebrity news site TMZ shortly after he resigned showed him so frail that his skin looked nearly translucent. He looked like he'd aged decades since his last public appearance at the Cupertino City Council. Still, no one could have anticipated his passing so soon after. Most employees found out at the same time as the public. Apple's executive team was devastated.

Flags were flown at half-staff at Microsoft and Walt Disney World as well as at Apple. A photo of Jobs was displayed on the big screen at NASDAQ in New York, where the ticker normally ran. Apple's shares plunged 7 percent in after-hours trading.

Tributes poured forth from Gates, from Facebook cofounder Mark Zuckerberg, even from Barack Obama. The president called Jobs "among the greatest of American innovators—brave enough to think differently, bold enough to believe he could change the world, and talented enough to do it."

Twitter and Facebook also lit up with the news as soon as it was announced. Teary-eyed fans flocked to Steve Jobs's home and Apple's offices to pay their respects. They took photos and left behind flowers, candles, handmade cards, and other mementos at makeshift shrines. One fan showed up at headquarters in a Scottish kilt and bagpipe and played "Amazing Grace." Some left apples with a bite taken out of them.

In Apple Stores in Mexico City and Prague, computers displayed a black-and-white photo of Jobs with his name and the years he had lived, "1955–2011." A Beijing store switched off the large Apple logo that was usually lit up. Fans left stickies with messages at every location.

Thank you Steve for transforming our lives, said one at San Francisco's flagship location.

Thank you and go well, said another.

In Japan and San Francisco, fans held a vigil, holding up iPads and iPhones with an image of a candle.

6

Ghost and Cipher

His ghost loomed everywhere. Obituaries blanketed the front pages of newspapers and websites. TV stations ran lengthy segments glorifying how he had changed the world. Essays by anyone who had been touched by him circulated on the Internet. Jobs's former software chief Avie Tevanian posted a remembrance on his Facebook page about Jobs's bachelor party. Only Tevanian and another friend had shown up because everyone else had been afraid of being in a social situation with him. Even those with whom he had toxic run-ins wrote glowingly about him. Gizmodo's editor in chief, Brian Lam, expressed his regrets over the blog's handling of the iPhone 4 prototype in a eulogy titled "Steve Jobs Was Always Kind to Me (Or, Regrets of an Asshole)."

Recounting how he had forced Jobs to send a letter formally claiming the device, Lam wrote, "If I could do it again, I'd do the first story about the phone again. But I probably would have given the phone back without asking for the letter. And I would have done the story about the engineer who lost it with more compassion and without naming him. Steve said we'd had our fun and we had the first story but we were being greedy. And he was right. We were. It was sore winning. And we were also being short sighted." Lam confessed that he'd had moments when he'd wished they had never found the phone at all.

Though a handful of negative pieces recalled Jobs's tyranny, the vast majority of the coverage was reverential.

In New York, Simon & Schuster rushed out Isaacson's biography a month early. Jobs had no control over the book's content, but he had argued hard over the cover. In one of the initial versions, the publisher had proposed a cover with the Apple logo and Jobs's picture.

The title was "iSteve." Jobs had been so infuriated that he threatened to withdraw his cooperation.

"This is the ugliest fucking cover. It sucks!" he had yelled at Isaacson. "You have no taste. I don't even want to deal with you anymore. The only way I'll deal with you again is if you let me have some say on the cover."

Isaacson agreed to allow him input. As it turned out, he would have needed Jobs's approval anyway because Apple owned the rights to every portrait of him that was any good.

A few months before Jobs died, the two exchanged endless emails about the photo and the font that would grace the cover. Isaacson persuaded Jobs to go with a *Fortune* magazine photo from 2006 in which the CEO stares out intensely through his round glasses with a hint of mischief. When celebrity photographer Albert Watson had shot it, he had asked Jobs to look 95 percent at the camera while thinking about the next project he had on the table.

Jobs won the argument to use a black-and-white version on the basis that he was "a black-and-white type of guy." Isaacson also acquiesced to Jobs's demand to print the title in Helvetica, a sans serif typeface that Apple had used in the past for corporate materials, but he refused to make the title, *Steve Jobs*, gray. Isaacson felt strongly that the title should be printed in black with his own name in gray.

"They're not reading Walter Isaacson's take on Steve Jobs," Isaacson had told him. "They're reading Steve Jobs as much as I can convey by getting out of the way."

One idea that had been floated by Simon & Schuster was to release the book with no title on the cover—publishing's version of the White Album. But Jobs rejected it, saying that it would appear arrogant. In the end, they settled on a sleek, elegant, and simple cover, much in the style of Apple's products.

When Jobs died, Apple chose the iconic image as the tribute photo on its home page. Both the photo and its effect were so quintessentially Jobsian that his friends and colleagues marveled at how the late CEO seemed to have orchestrated the narrative from beyond the grave.

Even the rituals of remembrances unfolded as though Jobs had staged them himself. A memorial service on a Sunday evening at

Stanford University was organized by his longtime event planner. The guest list read like a Who's Who of notables in Jobs's life—Bill Gates, Larry Page, Michael Dell, Rupert Murdoch, and the Clinton family, among others. Pixar was represented by John Lasseter, who made Jobs's favorite movie, *Toy Story*, and actor Tim Allen, the voice behind the character Buzz Lightyear.

Folksinger Joan Baez, Jobs's onetime girlfriend, sang "Swing Low, Sweet Chariot." Bono performed Bob Dylan's "Every Grain of Sand."

"Don't have the inclination to look back on any mistake," Bono sang. "Like Cain, I now behold this chain of events that I must break."

Yo-Yo Ma brought his cello at the personal request of Jobs before his death. Steph Adams, Jobs's event planner, roamed around with a headset, making sure nothing was amiss. Afterward, the attendees moved to a sculpture garden nearby and enjoyed hors d'oeuvres and wine in the brisk evening air. The event was tasteful, perfect, and unforgettable.

Jobs was gone, but not gone. Somehow he had transcended death to obsess over the launch of one last product: his own legacy.

Tim Cook was at the service but did not make a speech. Unreadable as always, the new CEO kept a relatively low profile. People saw him that evening but gave him little thought.

Even as he took control of Apple's sprawling empire, Cook could not escape his boss's shadow. The question was, how would Cook leave that shadow behind? How could anyone compete with a visionary so brilliant and unforgettable that not even death could make him go away? He was now officially larger-than-life.

The genius trap had been set long before for Jobs's successor.

Jobs was a superstar without whom Apple would not exist. The entire company had been defined by him for more than a decade, and few decisions were made without his input. Design, product development, and marketing strategies hinged on his opinions and tastes, while his executive team was handpicked based on their compatibility with Jobs and their ability to compensate for his weaknesses and bolster his strengths. Though Apple's accomplishments were not Jobs's alone, he had taken credit for most of them, which fed into his legend as a once-in-a-lifetime visionary. One employee even owned a car with the vanity plate: WWSJD—What Would Steve Jobs Do?

The legend of resurrection and triumph, carefully burnished and nurtured, had worked in Apple's favor when he had been at the helm. But without him, the empire was lost. To prevent it from collapsing, Cook would have to construct a different narrative quickly. To add to the challenge, Apple was a global corporation now, and it had to find a way to sustain itself as such while continuing with the transformative innovations for which Jobs had been so famous. In all likelihood, Jobs himself would have struggled to manage the company's bigger scale. Even before his death, he had demonstrated a reluctance to accept that Apple was no longer a nimble start-up. But his passing made managing the company that much more difficult. The next CEO didn't have the quasi-religious authority that Jobs had radiated. His every decision would be examined by current and former employees and executives, investors, the media, and Apple's consumers. He would also have to contend with the sky-high expectations that Jobs had conditioned the public to have for Apple.

Cook was a seasoned businessman and arguably a better manager than Jobs. He was organized, prepared, and was more realistic about the burdens of a company of Apple's current size. Many even considered him to be a genius in his own right. But no one could beat Jobs at being Steve Jobs, especially Cook, who was his polar opposite.

Together, the two men had struck the perfect balance. If Jobs was the star, then Cook was the stage manager. If Jobs was idealistic, Cook was practical. "Steve and Tim were very different people," one executive who worked closely with them said, adding that Cook would haggle over a nickel to drive profits whereas Jobs would spend a nickel to make people happy. "Tim was very much, 'Hey, we can't do that,' and Steve was like, 'We have to!'"

Jobs trusted Cook because he lacked any external demonstration of ego. Even when Cook disagreed with Jobs, he had mostly done so privately and cleverly. Rarely did he directly oppose Jobs. Instead, he proposed possible alternative solutions in a conciliatory manner, often with data to support him. "He was super smart about handling Steve," said Mike Slade, who worked with Cook as an advisor in the late 1990s.

Without Jobs, however, Cook had no counterweight to his dogged pragmatism. Who would provide the creative sparks?

For more than a decade, Cook had served his master. He set up and managed a world-leading supply chain system that could mass-produce Apple's groundbreaking products perfectly, profitably, and quickly. Whatever Jobs wanted, he made happen. When executive recruiters had come calling with CEO positions at Dell and Motorola, Cook had politely turned them down. He loved Apple and had no plans to leave, he would say, adding that he was perfectly content to be second-in-command.

But underneath the demure denials were hints of grand ambition. One of Cook's favorite quotes was one from Abraham Lincoln. "I will prepare and someday my chance will come." It was this belief that had led him to study industrial engineering, join IBM, then Compaq, and finally Apple. He had prepared and waited with seemingly infinite patience. Now his moment had arrived.

Still, the succession was complicated by the fact that no one knew who Cook really was. The new CEO was a cipher, a blank slate. As far as anyone could tell, he had no close friends, never socialized, and rarely talked about his personal life. Even physically, he didn't stand out. He was tall and lean with broad shoulders, pale skin, and graying hair. He wore barely noticeable rimless glasses and favored button-down shirts and polos. The most distinguishable part of his attire was his Nike sneakers, chosen because he sat on that company's board.

When an interviewer asked him what kind of a leader he was, he dodged the answer, saying, "I'd let other people describe me."

By design, he remained a complete mystery.

Cook was full of contradictions. To some, he was a machine. To others, he was riveting. He could strike terror in the hearts of his subordinates, but he could also motivate them into toiling from dawn to midnight just for a word of praise.

No one considered Cook to be antisocial, but neither would they describe him as social. He was friendly and pleasant, and those with passing interactions saw him as a gentle southerner with an aura similar to Mr. Rogers. But he wasn't approachable.

"I'm a hugger and a kisser," said Gina Gloski, a college classmate, who had sat on the alumni council with Cook. "But I'd never feel comfortable giving Tim a hug or a kiss." When one of his high school teachers emailed to congratulate him on a big promotion at Apple, he didn't respond.

Over the years, colleagues had tried to engage him in personal conversations, with little success. He worked out at a different gym than the one on campus and didn't fraternize outside of work. Years before, when Apple was about to ship its movie editing software, iMovie, Jobs wanted the executive team to test it out by making home movies. Jobs's was about his kids, and hardware executive Jon Rubinstein's poked fun at the meetings he had to attend on his birthday. Cook made his about house hunting and how little one got for their money in Palo Alto in the late 1990s. While amusing, the movie revealed nothing about him.

Few people knew how close Cook was to his family. His colleagues were aware that he went home occasionally, but most of them had no idea that he called his mother every week.

"He calls every Sunday, no matter what, no matter where he's at," Cook's father had told a television reporter in 2009 as he sat in a La-Z-Boy chair. "Europe, Asia, no matter where he's at, he calls his mother every Sunday. He don't miss a one."

After the interview, his father handed the TV reporter a bag of Satsuma oranges that he had picked from his backyard.

His parents were still living in Robertsdale, Alabama, the small, predominantly white rural town near the Gulf of Mexico where they had raised Cook. Their brick ranch house stood on a dead-end road in a pleasant middle-class neighborhood. Neatly trimmed evergreen bushes and a few flowers adorned the front side. The living room was modest with family photos and memorabilia, including track-and-field trophies that Cook had won as a child. A pair of Korean dolls wearing *hanbok*s, the national costume, sat on the mantel as a souvenir from one of Cook's many business trips to Seoul. The only fancy item was a flat-screen television on their wall that was connected to a DirecTV satellite dish. There was no computer in sight. His parents considered themselves to be too old to learn how to use one.

The Cooks were proud of all three of their sons, but their middle child, Tim, was the town's biggest success story even before he was named CEO. In December 2008, shortly before Jobs went on leave for the second time, Donald and Geraldine Cook walked into the offices of their local newspaper, the *Independent*, and offered an interview about their son.

"I am proud he has done so well," his mother told the reporter assigned to conduct the impromptu interview. "Nobody helped him."

"He's a workaholic," added her husband. They talked affectionately about how he stopped at the gym every morning to work out.

When Apple's media relations department found out that the *Independent* was planning to publish a story about the interview, they asked the editor not to publish it, to preserve Cook's privacy. Recognizing that they had a scoop on Robertsdale's most famous son, the newspaper refused. But to mollify the corporate giant, it kept the piece off the front page.

Most reporters who covered Silicon Valley had never interviewed Cook. At his request, his parents soon stopped speaking to journalists as well. When their son was named CEO, they referred all press inquiries to the family's pastor.

For a man who craved invisibility, Cook could not have had a more perfect childhood. His southern upbringing was so ordinary that the guardians of secrecy in Apple's public relations department could have scripted it. Those who would later turn to his early days for clues would find little to crack the code. That his unknowability had taken root so young only added to his aura.

During his early years, Cook's family lived in Pensacola, Florida, on the coast of the Gulf of Mexico. Cook's father worked at a shipyard for Alabama Drydock, rising to assistant general foreman. His mother was a homemaker. Eventually they moved to Robertsdale, a small town thirty-eight miles northwest of the city. Donald Cook later explained in an interview that he and his wife had chosen Robertsdale for its schools.

Though he did not disclose the exact date, the family's move appears to have been in the early seventies, a time when many families were fleeing Pensacola due to escalating racial strife over desegregation in the public schools. Cook later spoke of the terrible discrimination he had witnessed growing up. In historically white schools, teachers still routinely referred to the Civil War as the War of Northern Aggression. At Escambia High School, the old traditions died especially hard. In 1972, three years after integration became the law, the school band infuriated black students by flying the Confederate flag and playing "Dixie." The tensions, which erupted in a riot at an Escambia football game, continued for the next several years.

Terrified at the prospect of further violence, many families began moving out of the school district. At the time, Tim Cook was in middle school. His photo first appears in a Robertsdale school yearbook in 1972 at the end of his sixth-grade year. Robertsdale was quiet, stable, and so safe it bordered on boring. Outsiders found the place easy to forget because it was literally a stop on their way to the beach. Known as "Hub City" for its central location in Baldwin County, Robertsdale was the kind of town where people's social lives revolved around church, the Golden Bears high school football team, and seasonal events like the Shrimp Festival in nearby Gulf Shores. About the only thing that ruffled the town's calm was the advent of hurricane season. Living so close to the Gulf, the residents held their breath as they watched the barometer and studied the horizon. The closest call during Cook's adolescence was Hurricane Eloise, a storm that made landfall near Panama City in 1975 and passed just to the east of the town, tearing up the area with top winds of up to 155 miles per hour.

For the most part, life in Robertsdale followed a predictable routine. Christmas was marked with a parade and visits from Santa, sponsored by Delta Chi. Dewy girls competed in the Junior Miss pageant. Methodist youth groups climbed onto buses headed for Six Flags. The stylists at Emma's Hair Fashions sought advice at the hair show in Panama City; the Southern Life Insurance Company went to Pensacola for the district picnic. At summer's end, the residents slathered butter over ears of freshly harvested corn. In the fall, they lay in bed and listened for the soft clatter of pecans falling on roofs.

The *Independent* treated rattlesnake bites as big news, especially if the offending fangs came from the mouth of a five-foot-long diamondback. When the local grocery store, Morgan's Corner, was sold to Piggly Wiggly, the paper was all over it. A front-page story detailed the experiences of a priest visiting from his home parish in Ireland. The headline: STAND-IN PRIEST NOW LEARNING TO DRIVE ON RIGHT SIDE OF ROAD.

Inside the pages, columnists faithfully recorded Bible camps, surprise birthday parties, visits to great-grandmothers in Georgia, spaghetti dinners, updates on residents recovering from surgery, backyard barbecues enjoyed by out-of-town guests, as well as the names of the ladies who attended a coffee for Mrs. Fob James at the Lake Forest Country Club. No item was too small.

"Thursday's Bridge Club," one story reported, "was held in Mrs. Juanita Freeman's home with Iris Malone winning high and Corina White winning bingo."

The *Independent*'s voluminous coverage was well-intentioned, even sweet. But growing up in this environment, Cook could be forgiven if he developed an aversion to the public reporting of private life. Not that he had a problem with newspapers. As a teenager, he made extra money delivering for the *Press-Register*. Along with his mother, he also worked part-time at the local pharmacy, Lee Drugs, where he waited on customers and stocked shelves. Friends and teachers remembered him being confident and funny.

"He wasn't quiet at all," said high school classmate Susan Baker. "I don't know a soul who didn't like him."

Where Cook really excelled was in the classroom. The public school taught kindergarten through twelfth grade, all in one U-shaped, red-brick building. The annual talent show was dubbed Bunny's Good Time Hour after Barbara Davis, a popular teacher whose nickname was Bunny. Cook never participated.

From a young age, he showed a drive to succeed. Exhibiting an early aptitude for analytics, his favorite subjects were algebra, geometry, and trigonometry.

"He was a good problem solver," recalled Davis, who taught Cook math for three years. "He would stick with something until he got it."

He was conscientious even in subjects like typing. His teacher would use a sheet of paper to cover the fingers of students who looked at the keys too much, but she never had to do that with Cook. "He was so concerned about his grades," said Dolores Teem. "I just told him, 'Do what I tell you and you're going to do fine.' "

From seventh through twelfth grade, he was voted "most studious." He represented his town at Boys State, an American Legion mock legislature program, and won the Optimist Oratorical contest on the theme "Give Me Your Hand." In his junior year, he won a contest organized by the Alabama Rural Electric Association by writing an essay on the topic "Rural Electric Cooperatives—Challengers of Yesterday, Today, Tomorrow." First prize was an all-expense paid trip to Washington, D.C., where he attended banquets, visited sites like Arlington National Cemetery, and heard President Jimmy Carter speak at the White House. Had he won second prize, he would have received a savings bond.

"He's the kind of fella who don't believe in giving up on nothing," Cook's father told the local CBS affiliate, WKRG. "He's a go-getter."

Outside of class, Cook played the trombone in the school band, which performed at football games, parades, and school dances and was considered the town's mainstay of musical entertainment. The band practiced every Tuesday and Thursday after school. Wednesdays were reserved for families going to church. Cook also worked on the *Robala* yearbook staff, where he served as business manager. Davis, who was also the yearbook advisor, chose him for the job because she knew he was meticulous and good with numbers.

Like so many high school yearbooks, then and now, the *Robala* steered clear of controversy and ambiguity with the slightest hint of teen angst. Nearly every page radiated a dedication to portraying the school in the most idealized light possible.

Scanning the volumes from Cook's high school years takes the reader on a forced march through the fashion wasteland of the mid-seventies: helmet hair, leisure suits, bell bottoms, even silk screen shirts with psychedelic prints. Like the rest of the nation's youth, the students of Robertsdale High cried out for a mass intervention. The year after *Star Wars* hit theaters, the book showed one girl sporting Princess Leia's signature cinnamon buns, seemingly without a trace of irony. Cook was not immune to the ravages of the age. His photos show him smiling awkwardly beneath giant waves of hair, his gangly frame squeezed into printed button-down shirts and loud checkered slacks. In his junior yearbook, a photo showed him listening to a tape with a pair of large headphones next to a classmate on an electric typewriter. "Teresa and Tim," the caption read, "are using two of the modern ways to help study." In another, Cook stood next to his friend Lisa as her escort in the Homecoming Court.

Decades later, Robertsdale's mayor, Charles Murphy, would speak proudly of their native son. "It's a big honor for us, someone who was raised here, it just shows that through hard work and diligent effort what you can accomplish."

He added, however, that he had met Cook only once. When the executive returned home for a visit, he did so quietly. A cashier at Lee Drugs said his parents still patronized the store, but she had never met their son. "I've seen him on television," she said. "He favors his daddy."

Cook had always wanted to be an engineer, and after graduating in 1978, he enrolled in the industrial engineering program at Auburn University, a school whose football team he had been a fan of since he was a child. Where he came from, you were devoted to either the Auburn Tigers or the University of Alabama's Crimson Tide. Even back then, Cook had an affinity for underdogs, and the Tigers qualified.

The future Apple executive couldn't have picked a more suitable major to start on the road that would eventually lead him to Cupertino. Industrial engineers optimize complex processes or systems. In contrast to other types of engineers who focus on specific technical problems, industrial engineers look at the big picture to eliminate waste and figure out the most efficient use of resources. This training makes them exceptionally qualified to move into management. Former Chrysler CEO Lee Iacocca, Wal-Mart Stores CEO Mike Duke, and former United Parcel Service CEO Michael Eskew all came from industrial engineering backgrounds.

Cook did well in college but didn't stand out. He tended to sit in the middle of the classroom and rarely asked questions or visited professors during office hours.

"I had no visions of grandeur for him," admitted Saeed Maghsoodloo, one of his professors. He added that it was difficult to predict a student's success at such a young age. "He quietly did everything by himself, and as far as I remember, he was at least a solid B-plus or A-minus student."

Some of his teachers recalled his innate talent in one particular area. He was faster than anyone else at identifying problems in case studies.

"He could cut through all the junk and get down to the gist of the problem very quickly," said another professor, Robert Bulfin. "I suspect that has served him very well."

Auburn had a reputation for being tight-knit. Students mostly came from Alabama or neighboring southern states, and their shared backgrounds, coupled with the school's relatively isolated location and strong football culture, bound them together. The school referred to its community often as the "Auburn family." Its guiding principle was the Auburn Creed. "I believe that this is a practical world and that I can count only on what I earn," it began. "Therefore, I believe in work, hard work. I believe in education, which gives me the knowledge to work wisely and trains my mind and my hands to work skillfully. I believe in honesty and truthfulness, without which I cannot win the respect and confidence of my fellow men."

The vow, written by Auburn's first football coach, George Petrie, was brutally pragmatic, especially its first two sentences, which resonated strongly with Cook. Nearly thirty years later, he would tell students in a commencement address at Auburn that for as long as he could remember, the line "I believe in work, hard work" had formed one of his core beliefs.

"Though the sentiment is a simple one, there's tremendous dignity and wisdom in these words and they have stood the test of time," Cook had said. "Those who try to achieve success without hard work ultimately deceive themselves, or worse, deceive others."

This work ethic would come to define him as he built his career.

Cook's experience at Auburn also helped him in a more tangible way. Partially to offset his tuition, Cook enrolled in a cooperative education program in which he spent part of his time working at Reynolds Aluminum in Richmond, Virginia. While he was there, the company laid off most of its employees, giving him the opportunity to fill in and run the company alongside the president. It was his first taste of corporate management. A later stint at the Scott Paper Company gave him insight into another traditional industry. In his senior year Cook was nominated as outstanding industrial engineering graduate. "I don't deserve this," he said then. "There are any number of people who deserve this more than me."

An IBM recruiter who happened to be present hired him. The computer giant was one of three potential employers Cook had been considering. The other two had been Andersen Consulting and General Electric.

"The truth is, I'd never thought much about computers. It just sort of happened," he recalled years later. "Would things have turned out different if that hadn't happened? I don't know. But I *do* know that there are only a very few things in life that define you and that was one of them for me."

Decades later, he would talk about how uncertain he was about his future at his graduation. When IBM sent him to Duke for an MBA, he had even drawn up a twenty-five-year plan as part of an exercise.

"Let's just say it wasn't worth the yellowed paper it was written on," he told graduating Auburn seniors in 2010. "Life has a habit of throwing you curveballs. Don't get me wrong—it's good to plan for the future, but if you're like me and you occasionally want to swing for the fences, you can't count on a predictable life."

Life under Jobs was a roller coaster, but Cook's operations fiefdom was orderly and disciplined. Cook knew every detail in every step of the operations processes. Weekly operations meetings could last five to six hours, as he ground through every single line item. Even a small miss of a couple of hundred units was examined closely. "Your numbers," he said flatly to one planner, "make me want to jump out that window over there."

In another meeting he wanted to know why a manager in Ireland wasn't teleconferencing in. "Where's Joe?" he asked as he slowly crushed a can of soda in his hand.

His subordinates quickly learned to plan rigorously for meetings with him almost as if they were studying for a school exam. They built financial models and prepared detailed budget estimates.

"If you ever spent money without it being fully approved all the way to the highest level," recalled one staffer who worked under him, "it was a serious fucking infraction."

Meetings with Cook could be terrifying. There was occasionally idle chitchat, but it was mostly nervous chatter. Cook, who exuded a Zen-like calm, didn't waste words.

"Talk about your numbers. Put your spreadsheet up," he'd say as he nursed a Mountain Dew at his side. It was a puzzle to his staff why he wasn't bouncing off the walls from the caffeine.

When Cook turned the spotlight on someone, he hammered him with questions until he was satisfied. "Why is that?" "What do you mean?" "I don't understand. Why are you not making it clear?" Bam, bam, bam.

He was known to ask the same exact question ten times in a row, but he also knew the power of silence. Cook could do more with a pause than Jobs ever could with a swear word. When he wanted something done, he would just stop talking and expect the other person to fill the void. When someone was unable to answer a question, he would sit there without a word while people stared at the table and shifted in their seats. The atmosphere was so intense and uncomfortable that it made everyone in the room want to back away. Cook, however, didn't move a finger as he focused his eyes on his target. As the person squirmed in his seat, the room would be dead still. Sometimes, an unperturbed Cook would take out an energy bar from his pocket while he waited for an answer. The silence would be broken by the crackling of a wrapper.

Regular attendees swore that his pulse never rose as he barbecued his victims. His people learned to never let him get to the third "Why?"

Cook had little tolerance for people who weren't doing what they were supposed to be doing, be they on his team or at one of the suppliers. "When we would work with companies, he'd always want to know who was the man in charge," recalled one longtime operations manager. "That was his big thing."

Even in Apple's unrelenting culture, Cook's meetings stood out as harsh. On one occasion, a manager from the hardware group who was sitting in was shocked to hear Cook say to an underling, "That number is wrong. Get out of here." Hardware meetings were tough, too, but participants were never treated like that.

Cook's quarterly reviews were especially torturous because Cook would grind through the minutiae as he categorized what worked and what didn't, using yellow Post-its. His managers, meanwhile, crossed their fingers in the hopes of emerging unscathed. "We're safe as long as we're not at the back of the pack," they would say to each other. Even as they quaked, however, they marveled at how effectively he used fear and controlled tension to get results. His tough approach commanded respect because he set the same high bar for himself.

Cook demonstrated the same level of austerity and discipline in his life as he did in his work. He woke up at 4:30 or 5 a.m. and hit the gym several times a week. He ate protein bars throughout the day and had simple meals like chicken and rice for lunch. His diet was so predictable that when he and hardware chief Jon Rubinstein were traveling together, Rubinstein could order for him. During one business dinner at MacArthur Park restaurant in Palo Alto, Rubinstein proved it. "He'll take the salad with the dressing on the side and bring him the fish with the sauce on the side," he said to the waiter.

As everyone laughed, Cook confirmed the order. "Yes, that's what I'll have."

Rubinstein ordered a full rack of ribs and a side of fries.

His stamina was so inhuman that some wondered if he was actually human. Cook worked twelve to fourteen hours a day but insisted he was not a workaholic. Prior to Monday's executive team meetings, he would hold prep meetings on Friday afternoons and recap meetings on Sunday evenings. He could fly to Asia, spend three days, fly back, arrive at 7 a.m. at the airport, and be in the office by 8:30, interrogating someone about some numbers. During one eighteen-hour flight from California to Singapore several years before, Cook had stayed awake the entire time, preparing for a review of Apple's Asian operations as he soared over the Pacific. When the plane landed at dawn, he took a quick shower and headed into a twelve-hour meeting. By the end, the local executives were exhausted, but he wanted to keep going.

Cook was also relentlessly frugal. Though he could afford better, for many years he chose to live in a rental unit in a dingy ranch-style building with no air-conditioning. He said it reminded him of where he came from. When he finally purchased a house, it was a modest 2,400-square-foot home, built on a half lot with a single parking spot. He bought underwear at Nordstrom's half-yearly sale, and his first sports car was a used Porsche Boxster, an entry-level sports car that enthusiasts called the "poor man's Porsche." When he mentioned his new acquisition at a staff meeting, one of his more outspoken managers expressed surprise.

"Tim, what the hell? You can afford any car, anything, and you bought a used Porsche Boxster?"

"No, you know . . . ," Cook had said, not really answering.

Even his hobbies were hard-core—cycling and rock climbing. During vacations, he never ventured far. Two of his favorite places to visit were Yosemite and Zion. He was once spotted at Canyon Ranch Resort in Arizona, where he kept to himself, often dining alone reading on his iPad. Once when Joe O'Sullivan was recounting tales from a vacation in India he had just taken, Cook wistfully commented, "I'd love to do that."

"Well, why don't you?" O'Sullivan answered. "You're only young once, you know."

Cook's response: "I will sometime."

With his colleagues, Cook generally kept a professional distance. When they invited him to join them for social events, he wouldn't say no, but he never said yes.

"That's a good idea," he'd say noncommittally.

Cook also was a lifelong bachelor who never mentioned friends or romantic interests. Some blogs speculated that he was gay, and *Out* magazine ranked him as number one in its Annual Power 50 above Ellen DeGeneres for three straight years, but Cook himself never confirmed or denied the fact. In any case, his sexual orientation was irrelevant. Cook had little personal time. He might as well have been married to Apple.

Cook occasionally showed flashes of his old humor as he revealed the humble Alabama boy inside. Once when he was schlepping from one end of London's Heathrow Airport to the other to transfer planes during a business trip, his traveling companion was amused to hear him muttering under his breath, "Now I understand why people have private jets." Later when they were finally seated in the airport lounge, he took a package of cookies, looked at it, and wanted to know why they were called digestives. "Are these something I need to be careful of?" he asked. When he was reassured that it was just a type of British cookie, he joked, "With a name like digestives, I didn't know what I was going to be eating."

Even those who were witness to some of those moments, however, didn't know much about him aside from the fact that he was a die-hard Auburn Tigers football fan and admired Lance Armstrong, at least until the athlete admitted to doping. Cook's office was decorated with Auburn paraphernalia, and he reportedly modeled his close-cropped haircut after the seven-time Tour de France winner. One of his favorite quotes by the cyclist was "I don't like to lose. I just despise it." Cook had put up a slide of the phrase in an operations meeting once.

"In business, as in sports, the vast majority of victories are determined before the beginning of the game," he had said in his commencement speech at Auburn. "We rarely control the timing of opportunities, but we can control our preparation."

Like Jobs, Cook claimed to be a Bob Dylan fan. Unlike Jobs, it was difficult to imagine him cuing up "Subterranean Homesick Blues" on his iPod. Though he had grown up in a conservative part of the country, he considered Robert Kennedy and Martin Luther King Jr. to be among his heroes. Photos of both men hung in his office, and he told interviewers that he took inspiration from their example. "I always felt that Bobby Kennedy and Martin Luther King did an incredible amount for the whole of the world," he told business school students at his MBA alma mater, Duke. "They didn't solve it because it's not solved today, but they moved things forward in a major way. . . . And they knowingly risked their lives for it."

On a different occasion, Cook said that he admired the way Kennedy had been comfortable standing in his brother's shadow, doing what he thought was right. He said he was "tormented" at times by thinking about how things might have changed had the civil rights activist become president. "He had a way of touching and relating to people of all walks of life," Cook had said. "He was one of the people who got close enough to the presidency who really loved people, who wanted to raise people up." Kennedy embodied everything that Cook strived to be—hardworking, principled, and charitable.

As tough as Cook was reputed to be, he was also generous. He gave away the frequent-flyer miles that he racked up as Christmas gifts, and he volunteered at a soup kitchen during the Thanksgiving holidays. At Auburn, he helped found an advisory committee of industry leaders for the Engineering Department, gave money for a scholarship, and contributed to a fund that helped the department buy new equipment and technology. In the past, he had participated in an annual two-day cycling event across Georgia to raise money for multiple sclerosis. Cook had been a supporter since he was misdiagnosed with the disease years before. "The doctor said, 'Mr. Cook, you've either had a stroke or you have MS.' Turned out, I didn't have either one. I was just lugging a lot of incredibly heavy luggage around," Cook told his alumni magazine in an interview, adding that the incident put the world in a different perspective.

"I've already done more than I ever thought I'd do. If you start fearing things, then you don't try anything new or different. If it doesn't work out, it's not the end of my world. I'll go ride my bike."

Jobs's departure presented a crisis for the company. But it was also an opportunity for Cook to assume control in a way he hadn't been permitted to in the past. In August 2011, a few months before Jobs passed away, Cook sent his first email as CEO to employees.

Team:

I am looking forward to the amazing opportunity of serving as CEO of the most innovative company in the world. Joining Apple was the best decision I've ever made and it's been the privilege of a lifetime to work for Apple and Steve for over 13 years. I share Steve's optimism for Apple's bright future.

Steve has been an incredible leader and mentor to me, as well as to the entire executive team and our amazing employees. We are really looking forward to Steve's ongoing guidance and inspiration as our Chairman.

I want you to be confident that Apple is not going to change. I cherish and celebrate Apple's unique principles and values. Steve built a company and culture that is unlike any other in the world and we are going to stay true to that—it is in our DNA. We are going to continue to make the best products in the world that delight our customers and make our employees incredibly proud of what they do.

I love Apple and I am looking forward to diving into my new role. All of the incredible support from the Board, the executive team and many of you has been inspiring. I am confident our best years lie ahead of us and that together we will continue to make Apple the magical place that it is.

Tim

Apple's employees rallied around Cook in the days following, but privately there was considerable anxiety about how Apple might be run. Employees in departments that had little to do with Cook until then worried about how their jobs might change. The operations team, familiar with his tough management style, worried about life becoming even more intense.

In his first days as CEO, Cook took two immediate actions. First, he promoted Eddy Cue, Apple's enormously popular vice president for Internet services. Cue had started out as an intern. In one version of a story that he told everyone, he was plucked out of the IT department by Jobs during a meeting in which he had dared to voice an opinion about the topic at hand. When Jobs looked at him and told him to shut up, an undeterred Cue spoke up again, causing Jobs to throw a pen at his forehead. Cue, who by then figured he had nothing to lose, braced himself and offered up his opinion for a third time. This time, he won Jobs's approval.

From that moment on, Cue became Jobs's guy, managing the iTunes group, and then eventually all of Internet services. He was Jobs's deal maker as well, leading the initial talks with AT&T when Apple was launching the iPhone and negotiating content deals with the music labels, movie studios, book publishers, and media companies. Cue was the face of Apple for its partners. He wasn't a pushover, but he was accessible, attentive, and always willing to listen. He was also a genuinely nice guy who loved fast cars and religiously followed his alma mater Duke University's basketball team.

Without him, Apple would have found it more difficult to provide the breadth of Internet services and content that the company offered.

For all of Cue's importance, Jobs had never promoted him beyond vice president. Some thought that Jobs didn't consider services to be a core part of Apple, worthy of a senior vice president status, but Cue had been welcomed inside the inner circle for years. Everyone in the company was united in their belief that he should have been promoted years ago. Cook's appointment of Cue as senior vice president may have been obvious and possibly even pre-planned, but the decision generated goodwill inside and outside the company at a crucial time when Cook was still forming his public image. He also had turned an important Jobs loyalist into an ally on his executive team.

The second decision Cook made was to start a charity program. He announced that Apple would match charitable donations of up to ten thousand dollars, dollar for dollar annually. This, too, was embraced. The lack of a corporate matching program had been a sore point for many employees. Jobs didn't believe in philanthropy because he thought money was only a temporary solution to solving the world's problems. He thought matching programs were particularly ineffective because the contributions would never amount to enough to make a difference. Some of his friends believed that Jobs would have taken up some causes once he had more time, but Jobs used to say that he was contributing to society more meaningfully by building a good company and creating jobs.

"I'd rather make things and do things to change things," he had said at a Top 100 meeting.

Cook believed firmly in charity. "My objective—one day—is to totally help others," he said. "To me, that's real success, when you can say, 'I don't need it anymore. I'm going to do something else.' "

The moves generated goodwill and signaled a shift to a more benevolent regime.

Within a few months after Cook took over as CEO, employees began noticing other changes. Though still shuttered to the outside eye, the company felt more open internally as the new CEO communicated with them more frequently via emails and Town Hall meetings. Unlike Jobs, who always ate lunch with Jonathan Ive, Cook went to the cafeteria and introduced himself to employees he didn't know, asking if he could join them. Without Jobs breathing down their necks, the atmosphere was more relaxed. Cook was a more traditional CEO, who infused Apple with a healthier work environment.

But even as Cook tried to forge his own path, he was still influenced by Jobs. Shortly after he became CEO, Cook told a confidant that he got up every morning reminding himself to just do the right thing and not think about what Steve would have done.

Outside of Apple, Jobs was everywhere as the biography by Isaacson hit bookstores. Peering out from the cover, Jobs's intense eyes seemed to be watching. The book, based on more than forty interviews with Jobs, became a springboard for a new dissection of his life. CBS's *60 Minutes* ran a long segment. Then came more books about Apple and Jobs and plans for at least two movies.

The company that Cook now steered had inspired the world, but it now faced an array of challenges, from increased regulatory scrutiny to patent warfare. Some of the empire's most confounding problems were developing on the other side of the world. In China, where so many toiled to keep the lines of production rolling, the costs of Apple's success were becoming impossible to ignore. Falling twelve stories in the middle of the night, a man named Sun Danyong had become an implicit challenge to the moral authority of Apple and Foxconn, one of the twenty-first century's tragic victims of the global economy.

Joy City

Sun Danyong had been groomed for martyrdom.

A quiet young man from a remote mountain village in the south-western province of Yunnan, he was known at home as Little Yong. As a boy he did well in school, but his farmer parents were so poor that he had to reuse the same notebook multiple times by erasing his old notes. When he was accepted into the prestigious Harbin Institute of Technology, he could only afford to get there by standing the entire way on the 2,800-mile train journey from home. At school, he was thinly clothed even in the dead of winter and only ordered half a dish at mealtimes. Upon graduation, Sun couldn't find an ideal job in his home province, so he took one with Foxconn in the southern city of Shenzhen near Hong Kong. He didn't like being so far from home, but he had been thrilled about working for a prestigious company that made products for Hewlett-Packard, Dell, and Apple.

"You two don't have to be so tired anymore," he had told his proud parents when he got the job. "It is time for you to enjoy life a little bit."

At Foxconn, Sun was placed in the product communications department, where he was entrusted with sending iPhone prototypes to Apple. The job wasn't difficult but carried much responsibility.

His nightmare unfolded in mid-2009, about a year after he started. That July, one of sixteen iPhone 4 prototypes in his possession went missing. A record of Sun's statement, his communications with his friends, an interview with Foxconn's head of security, and surveillance cameras on the premises have provided an extraordinary record of what happened next.

Members of the security staff arrived at Sun's department to investigate. According to a statement Sun made to investigators, he had picked iPhones from the production line and sealed them with a security strip on the previous Thursday. Because his manager hadn't issued the sending documents, he held on to the package until the next day. That was a mistake. According to company procedure, he should have placed it in the storage room for safekeeping.

When someone arrived to take receipt of the samples the following afternoon, Sun opened the carton so the person could check its contents. Sun then made a second mistake by stepping away. When he returned, he was told that a phone was missing. A concerned Sun checked the production line and the location where he had kept the box. He conducted another search over the weekend before reporting the missing phone to his manager the following Monday. As the last known person to handle the prototype, Sun came under immediate suspicion for stealing the device. China had an active black market industry, so Apple was extraordinarily strict about the handling of its products. Foxconn employed an army of guards who kept a careful eye on the factory lines.

But a colleague disputed part of Sun's account, saying that he had been present during the handover of the samples. The head of security, Gu Qinming, called Sun into the office again that night to explain the discrepancy. Gu later admitted that he became angry during the interrogation when Sun refused to deviate from his story. In an account that was corroborated by camera footage, Gu pulled Sun's right shoulder before sending him into a small room next door to think things over. Until that point, Sun had remained calm. When Gu informed him that the police would be questioning him the next day, he looked up nervously and rubbed his hands together. The cameras showed Sun finally leaving the office at 10:41 p.m.

It's unclear where he went next. He was receiving concerned texts from his girlfriend about his whereabouts, so he probably didn't go home. He more likely ended up at an Internet café because a little over an hour later, he logged on to the Chinese instant messaging service, Tencent QQ, to tell friends of his humiliation.

"I've never taken anything away," he wrote. "The sample lost has nothing to do with me."

When a friend asked if he had been framed, Sun mused, "I've

thought carefully. There are only two possibilities for the loss of the sample. One is that somebody unintentionally took it away before I packaged it. The other is that somebody intentionally took it away that evening or the next day."

The twenty-two-year-old also spoke about how he wasn't interested in filching the iPhone even though he was poor.

"It's such a headache. I don't know how to handle it," he said. "Even in the police station, let alone a company, force is not allowed according to law . . . how can you detain me in your place and use violence against me?"

As he typed, Sun grew increasingly agitated.

"It is my responsibility that a sample was lost. I can accept that and I feel very sorry. But I didn't fucking take it away!"

In message after message, his indignation poured forth.

"These fucking dogs! They are inhuman!"

In many of his texts, he lashed out at his managers and the company. More alarming was the way Sun's anger gave way to a grim foreboding.

"My colleagues, it is my last time to speak the truth. It is not me who took the N90 sample," he wrote at 1:11 a.m., using the iPhone 4's code name. When a friend suggested that they get together sometime, Sun replied, "I hope so."

At 1:19:18 a.m., he sent out a message directed at the security chief. "Dear Head Gu Qinming, I've been bullied by you to speechlessness. I hope you will get your retribution soon. Your ability to beat me and the new phone that's about to come out is only because Foxconn is so strong, not because you are strong."

After worrying about the student loans he hadn't paid off, Sun sounded eerily resigned as he wrote to his friend at 1:26:27 a.m.:

"I'm going to leave, Gao Ge, have a good rest. Thinking about not being bullied by others and not being a scapegoat tomorrow, I feel much better."

A short while later security cameras showed Sun heading toward his apartment building and looking in before he entered the elevator of the building next to it. He got off on the twelfth floor. Before the door closed, the CCTV in the elevator captured him, clad in a white shirt, jeans, and sneakers—standard work attire at Foxconn. Sun was standing on his toes and looking out one of the windows in the corridor.

Some time around then, he sent a text message to his girlfriend, still awake and anxiously waiting for him at home even though it was well past midnight. "Sweetheart, I'm sorry. Go home tomorrow! I got into trouble. Please don't tell my family and don't try to contact me. It is the first time that I have ever begged you. Promise me you won't! I really feel so sorry."

Cameras showed Sun jumping off the building at 3:33:52 a.m. When his father and older brother collected his remains, Foxconn handed them a promissory note for 360,000 yuan (about $57,810) in compensation. Sun had left behind very little: an induction cooker, a cheap laptop, and some books. Among them was a translated copy of Thomas Friedman's *The World Is Flat: A Brief History of the Twenty-First Century.*

When Sun's suicide became public, the Chinese media seized on it as evidence of the extreme pressures that workers suffered at the hands of Foxconn and Apple.

Apple issued a statement, expressing shock and regret. "We require that our suppliers treat all workers with dignity and respect," the company said. Foxconn also issued an apology, promising to do its best to help the family, review its management process, and provide counseling for employees. The company suspended the security chief Gu, who claimed he was branded a murderer on the Internet, and nothing more was ever said about the missing iPhone.

At first, the companies treated the incident as an isolated case, but it turned out to be the harbinger of an alarming pattern. Over the next two years, at least eighteen more suicides followed at Foxconn's factories, including that of Ma Xiangqian, the worker who leapt to his death after being demoted to cleaning toilets. Foxconn was slow to take the problem seriously because the number represented such a tiny fraction of the 681,000 workers under its employment. In terms of percentage, the suicides were about an eighth of China's overall rate and far below that of many countries, including the United States.

To the media, these statistics didn't matter. Given the high profiles of Apple and Foxconn, the suicides were headline news. Foxconn, officially called Hon Hai Precision Co., was a Taiwanese manufacturing behemoth that did business with most of the world's best-known electronics brands. In August 2007, shortly after the iPhone went on sale, Hon Hai's market capitalization was $43 billion, equal to that of its ten biggest global rivals combined. Apple was worth over $115 billion. But Hon Hai's staggering achievements were not reflected in the workers' wages. A November 2009 pay stub showed one worker's monthly earnings: 2,145 yuan, or about $320. That included $135 in basic salary, $68 from working 60.5 hours of weekday overtime, and $110 from 75 hours of weekend overtime. Many Chinese were already resentful about how the world took advantage of their cheap labor to make high-end products for overseas markets. It was easy for them to interpret the suicides as yet another sign of how they were being exploited.

By Chinese standards, Foxconn was far from a sweatshop. The conditions overall were a vast improvement for the workers, most of whom came from unimaginably poor families in remote areas of China. As low as their wages seemed, Foxconn workers were at the top of the industry's pay scale. In the Longhua Science & Technology Park, the company's walled facility in Shenzhen, the enormous campus encompassed dormitories and canteens, where the kitchen cooked three tons of pork and thirteen tons of rice a day for 369,000 workers. The company also periodically provided entertainment, inviting Taiwanese pop stars to give concerts or hosting a talent show in which the CEO himself was known to dance. Employment there was so sought after that hundreds, sometimes thousands, lined up daily at Foxconn's recruitment centers.

In 2010, Foxconn started a program called "Stars of Foxconn" to acknowledge top workers. The parents of the winners, who had been nominated by their peers, were invited to one of Foxconn's campuses at the company's expense to attend a special awards ceremony and banquet, where they were treated to dishes like braised Yellow River carp with baked noodles, tempura shrimp, and mushrooms in abalone sauce. The award recipients were also given a one-week trip to Taiwan, during which they visited Hon Hai's headquarters and saw sights like the famous Ali Mountain and Sun Moon Lake.

Factory work, however, was factory work: repetitive, stressful, and mind-numbing. Most of the workers were between sixteen and twenty-five, adrift from the support network that high schools and universities provided and unprepared to fend for themselves in the impersonal work environment. They'd been drawn to the city by dreams of a more glamorous life, and the long hours and limited options were crushing.

Foxconn ran the facilities as though they were military installations. Workers were foot soldiers, and CEO Terry Gou was their general. In Asia, Gou was a legend. In 1974, he had borrowed part of the initial investment of $7,500 from his mother to start a small company that made plastic channel knobs for black-and-white televisions. Through perseverance, charisma, and cunning, he turned the business into the preeminent contract manufacturer. His personal hero was the thirteenth-century Mongolian conqueror Genghis Khan, and he wore on his right wrist a beaded bracelet from the temple dedicated to the great emperor. Gou embraced a code of aphorisms, compiled in a book that read like a capitalist's version of Chairman Mao's *Little Red Book*. Some of the sayings could have been channeled from Sun Tzu, outlining precepts for the art of corporate war. Others struck a quasi-mystical note.

> *Leadership is a righteous dictatorship.*
> *Unless the sun ceases to rise, there are no aspirations that cannot be realized.*
> *Suffering is the identical twin of growth.*
> *The true heroes give their lives on the battlefield. They don't return for medals.*

The book contained 108 of these sayings.

> *The kung-fu masters of Shaolin Temple only attain their omnipotent skills after years of tedious, grueling, and mundane training.*
> *People who live in earthquake zones are often the most alert.*
> *Every rooster believes that it is him who summons the sun every morning.*

Gou had done whatever it took to succeed. He expected the same level of commitment and self-discipline from his people, be they executives, corporate employees, or factory drones. Longtime veterans and business partners talked about how the company lured new recruits into the grind even at its corporate headquarters in Taiwan. There was a saying among them: The first year was the honeymoon. In the second year, they worked you like a tiger. And in the third year, they worked you like a dog. The company sweetened the trap by offering stock options that didn't vest for several years.

No one questioned Gou's work ethic. The CEO immersed himself in the job with almost puritanical devotion. His Shenzhen office was little more than a functional, one-story bungalow with cement floors. On many nights after the suicides, he slept there in a makeshift bed hung with mosquito netting. To all who worked for him, the message was clear:

I am one of you. Your hardships are my hardships. Our shared sacrifices are our bond.

If any of the general's troops had ever dared impertinence, they might have noted that the egalitarian gestures were built upon a rich man's conceit. Out on the factory line, many of his minions earned barely enough to scrape by. Raking in the profits through their labors, Gou had become one of the wealthiest men in Taiwan. Yes, he sometimes chose to sleep in spartan settings. But he also slept in a luxury apartment in the most exclusive neighborhood of Taipei and a castle in the Czech Republic. His workers slept in dorms, as many as eight to ten in a room.

Motivating the hordes who toiled for Foxconn—and managing the sprawl of its operations—was a monumental challenge. The scale of it was difficult for outsiders to comprehend. Around the world, from China to Brazil, Foxconn employed more than a million people, a figure that dwarfed the manufacturing workforce of the entire American auto industry. Logistical headaches multiplied as the company mobilized and trained hundreds of thousands of employees—many teenagers—and then watched them closely to ensure that the smartphones, tablets, and computers were manufactured on time without defect. The challenge required staggering discipline.

In a Confucianist society where hierarchies were ingrained and people felt a strong sense of duty to their collective group, it wasn't surprising that pressure from the corporation would filter down through the ranks with increasing weight. As demand rose for Apple's products, Foxconn was pressed to keep pace with orders from the other side of the world. That meant workers had to work more overtime, often through weekends and holidays.

Foxconn couldn't hire people fast enough.

For a long time, the world had paid little attention to conditions inside the factories. Foxconn's customers, including Apple, were focused on product quality and manufacturing schedules. Consumers were too busy playing with their shiny new toys to wonder where the devices had come from.

The luxury of detachment had prevailed until the summer of 2006, when a British newspaper, the *Daily Mail*, published an article about the deplorable conditions under which Apple's iPods were being made. The piece alleged that workers at Foxconn's iPod factory in Shenzhen slept one hundred people to a room at the time, with only a few possessions and a bucket to wash their clothes. Workers, the story reported, toiled fifteen hours a day for a mere twenty-seven British pounds, or about fifty U.S. dollars, half the amount weavers earned in Liverpool and Manchester in 1805, adjusting for inflation. The newspaper said the workers' collective share of iPod sales was less than 2 percent.

"It's the nature of big business today," said one expert, "to exploit any opportunity that comes their way."

The revelations spread through the blogosphere. Though Apple declined to comment on the report, the shock waves reverberated internally. The allegations weren't a surprise to some in the operation group's rank and file who had worked with the factories and had been troubled by what they had seen: shantytowns rising outside the factory walls, people defecating into a nearby canal.

Apple had a strict code of conduct designed to safeguard human rights, worker health, safety, and the environment. Every supplier was required to sign it, but there was no process to make sure the rules were being followed. The whole point of outsourcing manufacturing was that you didn't have to deal with factory management issues.

As head of operations and Jobs's second-in-command at the time, Cook had been in a tricky position. While he and his senior team wanted decent factory working conditions, they also needed to keep putting pressure on suppliers. Even before the problems came up, Cook told his managers, "Be as relentless as you can without crossing an ethical line."

Keeping that balance was difficult. In meetings where production schedules were on the agenda, Cook himself often made it clear that he didn't really care how the suppliers made their deadlines and delivered the projected quotas.

"They aren't working on Sunday," he'd say. "Why aren't they working on Sunday?"

The scandal threatened to taint the aura of benevolent cool that Apple had so carefully cultivated. They were supposed to be the good guys. Not the evil empire.

Cook launched an investigation into the charges. An audit team of members from the human resources, legal, and operations groups inspected Foxconn's Shenzhen factory and interviewed one hundred workers—randomly selected line workers, supervisors, executives, security guards, and custodians. The team reviewed thousands of personnel files, payroll data, time cards, and security logs. A report issued several weeks later said the auditors collectively spent more than twelve hundred man hours and covered more than one million square feet of facilities.

To Apple's relief, the situation was not as awful as had been reported in the *Daily Mail*. The audit team found that Foxconn's employees worked longer hours than the sixty-hour work week permitted in Apple's code of conduct. Sometimes they also worked more than six consecutive days, breaking Apple's rule that they must have at least one day off each week. Two buildings that had been converted into dormitories during a period of rapid growth contained too many beds and lockers in one open space. In interviews, two employees told the team that Foxconn supervisors had disciplined them by making the employees stand at attention. But the team had found no evidence of forced labor. Most of Foxconn's dormitories, the team concluded, provided adequate living conditions.

The truth was likely somewhere in between the appalling accounts in the *Daily Mail* and the more optimistic report from Apple's team. The company had probably done what it could to ensure the accuracy of its investigation. Working hours and overtime reported in interviews were corroborated with line shift reports, badge-reader logs, and payroll records. But the team couldn't have been fully briefed on the situation by the workers. Part of the problem was that the employees would have had a much lower expectation about the working conditions than Westerners were accustomed to. Many of the line workers were unaware of their rights.

Chinese culture also didn't permit the level of frankness found in the United States. To Foxconn's employees, Apple officials were authority figures, important customers, and outsiders. Even if they promised no recriminations from Foxconn managers, Apple's best intentions were short of a guarantee. The more senior interviewees would have felt an ingrained responsibility to the organization to put on their best face. Employees may have aired some dissatisfaction, but they would not have been completely open.

For a while, Apple tried its best. It vowed to ensure compliance with its code of conduct and audit all of its final assembly suppliers by the end of the year. The company also retained an auditing and research organization and created a small internal team to manage supplier responsibility. That team took apart the code of conduct and rewrote it with more than a hundred specific requirements, including maximum occupancy for dorm rooms and the square footage of personal space each worker should have. Suppliers not only had to comply; they also had to prove they weren't in violation and show evidence that they had a management process in place to stay compliant. Where there were violations, Apple helped suppliers fix them.

The work required patience. Some suppliers, who threatened to raise prices to cover the extra costs associated with making improvements, had to be reminded about the code of conduct they had signed.

Initially, the team reported directly to the operations number two,

Jeff Williams, and met weekly with Cook. But as their priorities conflicted with the grail of production schedules, enforcement appeared to have lapsed.

During this period, the company did make one significant improvement in labor conditions—it put a stop to steep broker fees that many workers paid middlemen to get jobs at Foxconn. In a modern form of indentured servitude, people had been paying up to two years' salary plus transportation costs if they were traveling from afar. Apple ended the practice by imposing a hefty fine on suppliers who permitted this practice.

The public soon forgot about the factory workers. As far as most of them were concerned, iPods materialized magically in stores and on their doorsteps. Apple had long encouraged this perception by excluding details about the assembly process from its product narratives.

Still, in mid-2008, the world was reminded of the hands that built their treasured phones when photos of a young factory girl on an iPhone assembly line most likely in Shenzhen surfaced on an iPhone 3G purchased in England. Wearing a pink-and-white uniform and a cap tucked behind her ears, she could be seen bending close to the device and faintly smiling. Her gloved hands made the peace sign. The girl, who came to be called "iPhone Girl," became an overnight sensation when the photos were posted anonymously on Apple fan site MacRumors.

"It would appear that someone on the production line was having a bit of fun," wrote the person who had bought the phone and shared the images. "Has anyone else found this?"

The posting immediately drew more than 360 comments as readers speculated on the girl's age, circumstances, and working conditions. "That's nice that at least they have some fun in the drab of assembling technology," wrote one reader. "She looks about 12/13 to me! I don't think Stevo is going to be impressed at all. . . . It looks a little bit too much like child labour," another said. Some wondered whether she would get fired. The Chinese press called her "China's prettiest factory girl." Apple declined to comment, but a Foxconn spokeswoman told a reporter that workers testing the device had probably forgotten to delete the photos. According to an official, iPhone Girl had not been fired.

Around the same time, activists learned that dangerous metals from electronics manufacturing plants were polluting China's rivers and poisoning its people.

Given that half of the world's computers, cell phones, and cameras were made in that country, the damage was devastating. Wastewater containing copper, nickel, and chromium from plants making printed computer circuit boards was eating at people's bodies. The manufacture of batteries and other power supplies also emitted lead, causing lead poisoning and high blood pressure among people who lived nearby. The heavy-metal levels in some places were so high that sewage treatment plants couldn't treat the water adequately before it was used again. In Guangdong Province, where Shenzhen lay, the government had found that its rivers had discharged more than twelve thousand tons of heavy metals and arsenic into the sea in 2008.

The offenders were Chinese, but they were making products for the rest of the world. The companies working with them needed to be held responsible. After two years of research, a coalition formed by the well-respected Beijing environmental activist, Ma Jun, sent a letter in spring 2010 to the chief executives of twenty-nine companies. The notice informed them of the environmental and safety violations. The recipients included almost every top brand, from Japanese companies like Panasonic and Sony to Europe's Nokia, Siemens, and Philips, to U.S. companies like Intel, Motorola, Cisco, and Apple. No one got a pass. Not even Chinese companies like Haier, Lenovo, and BYD.

After a few months, all of the other companies except for one responded in varying degrees to the coalition's entreaties. Apple answered with a stony silence. The company refused to engage in any discussions even when Ma provided details about hazardous waste violations and toxic waste poisoning at a factory associated with iPhone touchscreen supplier Wintek. Apple's reasoning: Its long-term policy was to not disclose the supply chain. Though the company later acknowledged the case in an annual sustainability report, it told NGOs at the time that they could not find a link between Apple and the accused supplier.

This evasiveness drew Ma's attention.

"I could not accept that argument," he recalled. "If their operations—either manufacturing or sourcing—have affected other people's interests, no one should be allowed to say, 'I made a policy not to talk and you should leave me alone.'"

Ma took a closer look at the situation and found appalling environmental abuses among companies Ma called "suspected Apple suppliers." Particularly egregious was the use of a toxic substance called n-hexane. Suppliers were using the chemical because it did a better job than alcohol of cleaning parts such as touch screens, but n-hexane damaged the nerves, led to numb limbs, and dulled the sense of touch. One account, purportedly by a nineteen-year-old worker at a sub-subcontractor, was particularly heartbreaking.

"Around October 2009, it was a busy time for our factory, except that the hands of all of the workers felt numb when they washed them or put them in water," she wrote on a blog that Ma mentioned in a report. "December came and my coworkers were starting to shake when they walked, and I couldn't believe why this all was happening so suddenly. What's more, one after another of my coworkers would ask for leave and each time they left, it was for a long time. In January 2010, it happened to my own body, and by the time I realized it was already too late."

The writer, Xiao Zhan, described how she lost her strength and her ability to run. She was finally admitted to a hospital, where she had daily IV treatments, injections, and physiotherapy. When she was released nearly a year later, she was out of a job and still in need of treatment. She had already spent all the money she had saved from working overtime to make Apple's products. Instead of sending money to her family, she had become dependent on them.

At the Wintek-affiliated factory, workers who had gotten sick were encouraged to leave for a one-time payment of $12,000–$14,000 if they signed a form absolving the company of any responsibility for their welfare.

In early 2011, Ma's group released his detailed findings to the public. Titled "The Other Side of Apple," the report was accompanied by a shocking video that showed some of the sick workers discussing their symptoms in between clips of Steve Jobs touting the magnificence of Apple's products. Altogether Ma described eleven cases of supplier negligence at suspected Apple suppliers. They mentioned the suicides at Foxconn as well as a practice at another company, where female workers had to undo their belts and submit to a body inspection by a male guard before leaving the factory, to make sure they hadn't taken anything with them. The practice had come to light after a worker posted about it anonymously online:

"Watching a younger girl stand on the inspection platform with her pants suddenly falling down and run away as everyone laughed at her, my eyes filled with tears."

Apple may not have been directly responsible for these violations, but they were as culpable as the suppliers who created these conditions. "How has a company with such a poor record for such a long time been able to *successfully* maintain its near perfect, corporate, socially responsible image?" Ma's report asked, accusing Apple of hiding the true state of its supply chain behind its veil of secrecy. If true, the hypocrisy was particularly egregious considering that Apple's board member Al Gore championed environmental causes and had won a Nobel Peace Prize for his climate protection efforts.

Ma's group wasn't the first to address occupational safety issues and labor conditions, but his reports were particularly powerful because they were backed up by solid data. Ma himself was a former investigative journalist, whom *Time* magazine had named one of the top one hundred influential people in 2006. He was not only respected; he also knew how to present his story to make the biggest impact on the world.

Apple maintained its code of silence.

On May 20, 2011, not long after Ma's group released its damaging report, a polishing workshop at a new Foxconn factory in southwest China exploded when a buildup of aluminum dust in the air ducts ignited. Workers in a cafeteria nearby saw black smoke pouring out the shattered windows. Three workers died, and fifteen more were injured. The 250-acre facility had been expected to be the largest supplier of the iPad 2s. Built in just seventy-six days, the plant had begun operating before it was ready. Some of the workers were polishing the assembled devices in the cleaning stage at the end of the production process without adequate training. Wearing masks and earplugs, they worked feverishly without realizing the dangers of the aluminum dust that found its way into their hair and faces. When Wall Street analysts learned about the accident, their main concern was not the deaths of the three workers, or the possible poisoning of the others. What the analysts wanted to know was whether the explosion would delay iPad 2 production.

Ma continued to turn up the heat on Apple. Five months of investigations led to "The Other Side of Apple II." The forty-eight-page report detailed ten case studies of environmental violations at suspected Apple suppliers. In another video, Ma took a subtle swipe at consumers' ignorance about the conditions under which their products were made.

"Perhaps you love Apple. Perhaps you use Apple's products. Perhaps you long to own an Apple product. Are you aware of the story behind Apple?" it asked before showing viewers a milky-white channel of a river and copper- and nickel-infested waters near a suspected Apple supplier. Near another plant, a small boy complained of noxious odors, chest pains, and dizziness. Villagers knelt in front of the camera and clasped their hands as one woman, who already had her stomach removed due to cancer, clutched a plastic bottle with water from a nearby stream.

"We beg you, help us!" they pleaded. "Help us ordinary people!"

Finally Apple broke its silence. The company sent an email to Ma that acknowledged his report, pointed out a few errors where he had incorrectly identified a company as an Apple supplier, and asked for a phone call with the activist. Apple was up in arms over the report tarnishing the company's image. Cook's deputy, Jeff Williams, wanted to fly to Beijing immediately to talk to Ma to try to rein him in.

"Who is this person?" he wanted to know. Apple ultimately decided they needed to build a relationship with Ma rather than try to control him.

So began delicate conversations between Ma and Apple. Progress was slow, but at least they were talking. Apple demonstrated an interest in understanding Ma's group and its method of research. "Those meetings were candid meetings," said Ma. "Sometimes it was not easy. It was not always very comfortable." It was unclear how much of Apple's sluggishness had been affected by the leadership vacuum while Jobs was sick, but a few weeks after the former CEO's death, there was a breakthrough. By then, Ma had enlisted the help of the Natural Resources Defense Council, a prominent Washington, D.C.–based nonprofit group that works with corporations to protect the environment. In late October, Bill Frederick, an Apple operations executive, got in touch with NRDC's Linda Greer, an environmental toxicologist who ran the supply chain project. Frederick was a former IBM executive and a logistics expert. He had a reputation for being slightly more laid-back than Cook, but when he confronted a problem, he was just as single-minded.

Frederick admitted that environmental safety had been less of a focus but he told NRDC that the executives were starting to brainstorm some methods on how to keep better tabs on their suppliers. When Frederick asked for a meeting with NRDC, the organization agreed on the condition that Ma also be invited.

At the meeting, Ma, who spoke English fluently, did most of the talking, but Greer provided context. "I was the English-to-English translator," she joked. The meeting lasted for five hours. Some of the executives were defensive at times, and Apple continued to withhold details such as the discrepancies that they found between their audit and Ma's investigation. But they also acknowledged that they needed to be more transparent. At the end, the company agreed to do its own research to confirm Ma's findings and then to address whatever problems they found.

In Greer's assessment, Apple was working harder than most to rectify the situation. "They're in the top percentile," she said. "They're still not on top of the game, but none of them are." She hoped that Apple would become a role model for other companies. Ma also felt that the company was changing for the better.

That December, just as the negotiations between the company and the activists were starting, another explosion ripped through a factory operated by one of Apple's contract manufacturers.

The circumstances were eerily similar to the tragedy that had killed the Foxconn workers less than a year before. The explosion occurred inside another brand-new factory owned by a subsidiary called Pegatron that was gearing up for mass production of back-panel parts for the iPad 2. Just as in the earlier incident, aluminum dust had ignited in a section of the building where workers polished iPad cases. The explosion had been triggered during a test run, held on a weekend afternoon that December. Eager to win more of the corporate giant's business, Pegatron had been pushing hard to get the new production lines moving. The accident injured sixty-one workers, putting twenty-three of them in the hospital. Once again, the explosion demonstrated the complexity of the situation and the costs of Apple's relentless pressure on its overseas suppliers.

Finally the Western media began paying closer attention.

In early January 2012, a few weeks after the second explosion, public radio's *This American Life* aired an hour-long program about the abysmal working conditions of Apple's global supply chain. A long segment was reported and narrated by Mike Daisey, a performer best known for his one-man stage show, *The Agony and the Ecstasy of Steve Jobs*. In the public radio piece, titled *Mr. Daisey and the Apple Factory*, Daisey recounted his experiences touring Foxconn's factory in Shenzhen and speaking to underage and injured workers. His anecdotes were heartbreaking, especially when juxtaposed with Apple's rampant profits. In one scene, Daisey told the story of showing his iPad to a man whose hand had been mangled making the device.

"He's never actually seen one on, this thing that took his hand. I turn it on, unlock the screen, and pass it to him. He takes it. The icons flare into view. And he strokes the screen with his ruined hand, and the icons slide back and forth. And he says something to Cathy, and Cathy says, 'He says it's a kind of magic.'"

Cathy was the name of Daisey's interpreter. Coming after news of the suicides and the factory explosions, the episode had a profound impact. The podcast version, was downloaded 822,000 times and streamed 206,000 times, making it the single most popular podcast in the show's history.

A few months later, *This American Life* retracted the story after discovering that Daisey had fabricated many of the details. There had been no underage workers and no man with a mangled hand. He had also lied about the number of factories he'd visited and the number of workers with whom he had spoken.

"I'm not going to say that I didn't take a few shortcuts in my passion to be heard," Daisey admitted to the show's host, Ira Glass. "My mistake, the mistake I truly regret, is that I had it on your show as journalism, and it's not journalism. It's theater."

Despite the fabrications, the momentum created by the original segment was unstoppable. A few weeks after Daisey's original program, the *New York Times* took aim at Apple in a damning series called iEconomy. The series opened with an article exploring the consequences of Apple's long-standing practice of moving its manufacturing overseas and profiting from cheap labor at the expense of America's jobless factory workers.

"Apple's an example of why it's so hard to create middle-class jobs in the U.S. now," Jared Bernstein, a former White House economic advisor, told the *Times*. "If it's the pinnacle of capitalism, we should be worried."

The article wasn't entirely fair to Apple. It ignored the fact that Apple would have surely gone out of business in the 1990s had it not started outsourcing its manufacturing to China. The *Times* reporters also didn't offer an economically feasible alternative to overseas outsourcing, a practice relied upon by many American corporations. For any major corporation that wanted to stay in business, the reality was that huge obstacles stood in the way of returning manufacturing to the United States. Part of the challenge was that the United States no longer had enough of the highly trained factory workers essential to churning out the millions of smartphones and laptops and other devices sold by Apple. Even if a massive army of skilled workers could be created, American labor costs were higher—which would translate into higher retail prices. For Apple to move manufacturing back to the States, consumers would have to be willing to pay a premium for their iPhones and iPads. Apple would lose customers to Samsung and other competitors manufacturing their products at cheaper costs.

Unlike Daisey's piece, the *Times* story was backed by exceptional reporting and gave frustrated Americans a target for their plight during a presidential election year when the economy was the center of conversation. The accusations wrecked the consumer-friendly image that Apple had carefully constructed. They also signaled a shifting tide in public sentiment.

Before the company could recover from that piece, the *New York Times* published a second story on the labor conditions and environmental practices at Chinese factories that make iPads. The article pointed out that more than half of the suppliers audited by Apple had violated at least one aspect of its code of conduct every year since 2007. Many of the violations echoed the content in Ma's reports, but it was fresh to Western readers.

"Apple never cared about anything other than increasing product quality and decreasing production cost," Li Mingqi, a former Foxconn factory manager, told the newspaper. "Workers' welfare has nothing to do with their interests."

The exposé, a direct hit from one of the world's most respected news organizations, set off alarms inside the mother ship. Less than six months after taking over, Cook needed to solidify his status as a leader, and the *Times* had raised questions no leader could ignore. Making matters worse, the timing was disastrous, at least from Apple's point of view. Many of the embarrassments detailed in the series had been focused on the production of iPads, and the *Times* had published the stories just as Apple was preparing to release its third generation iPad. That March, Cook scheduled a meeting in New York with the newspaper's editorial staff.

The showdown began pleasantly enough. Before Cook sat down, he chatted with executive editor Jill Abramson, telling her how he had been a faithful reader of the *Times* as a boy growing up in Alabama. "I loved the *New York Times*," Cook told her. "I'd run out and get it every morning."

For the first hour, editors and reporters plied the CEO with questions about upcoming products and his thoughts on topics like wearable computing. As the group broke for lunch, Charles Duhigg, the lead reporter of the iEconomy series, asked Cook what he thought of the stories. Cook had his back turned and was filling his plate. But he sat back down and opened fire. He accused Duhigg of relying on sources who weren't truthful. He protested the fairness of the entire series, arguing that the newspaper had been gunning for Apple from the start.

"I think you set out to write this story," Cook told the journalists. "There was nothing that we could have said that was going to change it."

The room went dead silent as Cook grew angrier. Though the intensity of his voice increased, the reporters in the room noted that Cook vented more quietly than Jobs, who had been a screamer. The new CEO's control made him all the more intimidating, especially when he emphasized one of his grievances by hitting the table with his fist.

For all of Cook's criticisms, the series prodded Apple to address the problems. In a statement similar to the one Apple issued after the 2006 *Daily Mail* article, the company announced that it was working this time with the nonprofit group the Fair Labor Association to audit final assembly suppliers, including the Foxconn factories that had been named in the newspaper article. The company said the association would review thousands of employees about working and living conditions, including health and safety, compensation, working hours, and communication with management.

Cook also sent a letter to employees. "We care about every worker in our worldwide supply chain. Any accident is deeply troubling, and any issue with working conditions is cause for concern," he wrote. "Any suggestion that we don't care is patently false and offensive to us."

In a meeting with senior executives, Cook vowed to rectify the situation. "Look," he said, "we're going to do the right thing no matter what the press says."

In March, Cook paid a visit to China to meet with political leaders and telecom executives and tour a new Foxconn factory.

In a shrewd public relations move, Apple did not announce his visit beforehand. Instead, Cook suddenly showed up at a Beijing Apple Store in a thirteen-story shopping center called Joy City that boasted one of the world's longest escalators. The choice of location for his surprise appearance was strategic. The shopping center wasn't Apple's most upscale location, but the store was popular with the masses. The press learned that Cook was in the country only after fans posted photos of themselves on the Internet, smiling next to the CEO, who was dressed casually in a navy Nike jacket with his glasses resting on top of his head.

"I met Tim Cook today in the Apple store in Joy City," one young woman tweeted excitedly in a widely circulated post on the Chinese Twitter-equivalent site, Sina Weibo. "I was so lucky to take a photo with him."

A couple of days later, Cook headed for Foxconn's newest plant in the north central city of Zhengzhou, where he donned a bright yellow factory cap and coat to tour the facilities. Apple hired a photographer to take publicity photos of him talking to a worker and smiling by an iPhone production line as he raised his hand at someone in greeting. That, too, made its way into publications around the world.

The highly public nature of the trip was rare for both Cook and Apple, and in the immediate aftermath, it seemed to pay off.

"China weighs more in Cook Era," the *Beijing News* daily declared. "In contrast to Jobs's policy to ignore China, Cook made China the destination of his first foreign trip as CEO."

Just by showing up and acknowledging the country's existence, Cook scored goodwill. But none of the coverage mentioned the reality that many of the problems he was responding to were of his own creation.

Years before, Jobs had given his second-in-command complete authority over the supply chain. Cook was the font of the pressure that his team put on suppliers and that the suppliers, in turn, put on workers. He was the one who told his team to do whatever it took to get the job done. Cook had no one else to blame. In one of the articles, the *New York Times* had recounted a famous story of how Apple had changed the iPhone screen from plastic to glass six weeks before launch. Internally, the story had been told and retold as an example of Apple's brilliance. But Cook had pressured Foxconn into making the change in an almost impossibly short window of time. The factory workers who made the iPhone had been called into work in the middle of the night. They started a twelve-hour shift to fit the screens into the frames after being given just a cup of tea and a biscuit.

8

Into the Fire

On an unseasonably warm day in London in May 2012, Jonathan Ive arrived at Buckingham Palace with his wife and twin sons, dressed formally in a black tailcoat and a cornflower-blue vest and tie. After being offered a glass of champagne, he took a seat in the ballroom. In recognition of his contributions to design, he was about to receive a knighthood.

The ceremony began with the lesser honors. Though the queen or the Prince of Wales usually conferred the awards, on this day the Princess Royal, the queen's eldest daughter, stood in for them. Shortly before his turn came, Ive rose to take his place.

"To receive the honor of knighthood and be a knight commander, civil division," a royal official announced, "Sir Jonathan Ive for services to design and to enterprise."

As Bach's *Double Violin Concerto* played softly behind him, Ive walked solemnly forward across the palace's red carpet. He stopped for a moment and gave a small bow before approaching Princess Anne, standing in her Royal Navy uniform. He knelt on the investiture stool with his right knee, and then the princess tapped his shoulders with a sword that had belonged to her grandfather, King George VI. When Ive stood up, she placed a medal around his neck, then chatted with him for a moment before moving on to the next honoree. Ive's hands, usually so expressive, stayed glued to his sides. But he talked to her animatedly about how often he returned to the United Kingdom. The princess spoke about her iPad.

He told reporters later that he was "both humbled and sincerely grateful" for the award. He described it as "absolutely thrilling."

"I'm extremely fortunate that I've found what I'd love to do, and that's essentially to draw and make stuff," Ive said. "To actually be able to spend all my time doing that, just that alone is fantastic, so then to get some recognition for that is . . . I don't know . . . It's a wonderful affirmation of the craft and the profession of design."

That afternoon, Ive changed into a more relaxed gray-blue suit with a dark blue-gray open collared shirt to attend a reception at the Royal Academy of Arts in commemoration of the queen's Diamond Jubilee and Britain's creative arts. He mingled with other guests, including Paul McCartney, Judi Dench, and Bono. Later on, when the queen awarded five special grants to up-and-coming young artists, Ive presented the award for design.

"I am the product of a very British design education," he told the audience as the queen stood near him in a white dress and a shimmery silver jacket. When the ceremony was over, a smiling Ive shook her hand and then stood behind her as she spoke to Bono.

Sir Jonathan had finally arrived.

Ive's knighthood was one more proof of his soaring prominence. Now that Steve Jobs was gone, and with him all his fiery brilliance, his protégé was widely seen as the most radically talented creative force remaining at Apple. If anyone could be relied upon to carry on the company's reputation for matchless bursts of elegant innovation, it was most likely to be Sir Jonathan.

It almost didn't matter if Ive truly was as indispensable as his reputation suggested. So many investors and stock analysts believed him to be irreplaceable that it became a self-fulfilling prophecy. At this delicate moment in the company's history, even the slightest whisper that Ive felt unappreciated could have been enough to send Apple's stock tumbling.

In the first months of his tenure, Tim Cook needed to deal with both the perception and the reality of the company's internal stability. Ive was key, yes, but so were the other top executives that the new CEO had inherited from his predecessor. After observing them during years of working together, Cook knew his senior team well enough to understand that they were all big personalities, fueled by egos and ambitions no longer tempered by the titanic presence of Jobs. A master of power dynamics, Jobs had managed his senior executives in a constructive way by giving each a separate domain of responsibility. On big decisions, though, he liked to throw them into the pit together, setting them against one another as they jockeyed to see who could come up with the cleverest solution to whatever challenges they faced. It was a harrowing but highly effective method that dared them to think harder and deeper, weeding through dozens of ideas until they finally arrived at a jewel of inspiration.

This approach had other benefits as well. When Jobs ruthlessly dismissed a proposal that the team felt strongly about, they relied on each other to change his mind. Jobs had transformed himself into a towering father figure, and all of his officers into his strong-willed but ultimately compliant children. Freud might have raised his eyebrows at such savage parenting. But the strategy worked, keeping his executives wary yet bonded to one another. Both competitors and collaborators, they had no choice but to take their cues from Jobs.

Now it was up to Cook to figure out how to keep the team working together effectively. Given the impenetrability of his personality, it remained unclear if he had the style and cunning to maintain their loyalty. Would Jobs's men stay at Apple and help him map the company's path into an increasingly volatile future? Now that their beloved father figure was gone, would they abandon their corporate home and take their vaunted talents elsewhere? Cook could not afford to let such questions linger. He had to establish his own rapport with Ive and his other top officers. Cook brought different strengths—and weaknesses—than Jobs to the equation. So did the others on the team.

The key, as Cook well knew, was to find a collective equilibrium. At one of the memorial services, he had talked about how his predecessor had modeled Apple after the Beatles.

"They were four guys who kept each other's kind of negative tendencies in check," Cook said, quoting Jobs. "They balanced each other, and the total was greater than the sum of the parts."

The reality, however, was that two superstars, John Lennon and Paul McCartney, had been given most of the credit for the band's genius. Once McCartney left, the rest of the band parted ways, showing the fragility of such high-powered collaborations.

Somehow, Cook had to forge a new dynamic with Ive and the others. If Apple was going to keep its juju, he needed to reconfigure the underlying currents of power so that the team orbited his personality, not their dead mentor. Together, they had to prove to the world that Apple could succeed without its biggest star.

To realize the magnitude of Cook's challenge, it was necessary to understand the personalities on the team.

The newest member—and Cook's first major hire—was John Browett, the senior vice president in charge of Apple's retail stores. The hiring of Browett had raised eyebrows because he had come from Dixons, a European technology retailer reminiscent of Best Buy. In contrast to Apple's clean, minimalist store environments, Dixons was a typical big-box retailer with brightly packaged products packed onto store shelves. At first glance, Browett seemed ideal because Cook had wanted someone with international experience to manage Apple's global retail business. Browett also seemed to care about customer service, one of the cornerstones of Apple's retail philosophy.

The Brit had a reputation for being a strategic thinker and a team player who was calm and focused—attributes that Cook was predisposed to appreciate. When Browett was running the British supermarket chain Tesco's online site, he had told a journalist, "I don't do lunch. I don't do conferences . . . there is too much to do."

But the newcomer had a great deal to prove. Browett was taking over one of the most successful retail operations in the world. Starting with just two locations in 2001, Apple had expanded the chain to include more than 400 stores, about a third of which were overseas. Annual retail sales per square foot was higher than at any other U.S. retailer, including Tiffany & Company. In fiscal 2011, the stores contributed $14 billion in sales. Browett had to find a way to grow Apple's retail profits even more while navigating an increasingly complex business. And he had to do so under the watch of a tightly knit executive team that had worked closely with Ron Johnson, his predecessor and the stores' original architect.

One of the most important and longest-standing members of

the team was Phil Schiller. As marketing chief to a marketing genius, the Boston native held one of the most difficult jobs at Apple. Jobs had constantly put him and his team under the microscope, second-guessing their every decision. Very little was good enough to pass muster. In some areas like advertising and branding, he had virtually no say. Jobs had managed those areas directly under an umbrella known as marketing communications. Throughout it all, Schiller survived and even flourished with a resilience that impressed everyone.

On the executive team, Schiller's contributions went far beyond just masterminding product launches and marketing strategies. He was a passionate participant in discussions about product development, helping to define products like the iPod, iPhone, and iPad. The navigation wheel on the iPod had been Schiller's idea, inspired by a Bang & Olufsen design. A technology junkie, he knew the ins and outs of every product's technical specifications and was respected for his extensive knowledge of market trends. Among those who dealt with him, however, he was a more controversial figure. Many loathed and feared him. Apple employees said Schiller was controlling and had a fiery temperament like Jobs, but compared to his longtime boss, he lacked charisma and gravitas.

Schiller had two nicknames. One was Dr. No, in reference to his Jobs-like habit of brusquely rejecting proposals. The other was Mini-Me, from the Austin Powers movies.

The name had been bestowed upon him because of the sidekick role he often played during Jobs's keynote presentations. One particularly memorable stunt during a 1999 Macworld event had Schiller leaping off a fifteen-foot platform to demonstrate how Apple's latest laptop computer could continue to transmit data wirelessly. As Jobs got the audience to join him in yelling, "Three. Two. One. Jump!," Schiller took off. He landed on a mattress as the accelerometer that was attached to the computer tracked his fall. "This is definitely one small step for man and one giant leap for wireless networking," he had joked beforehand. But the name stuck mostly because of his Jobs-like behavior. Even though people used the nickname snidely, Schiller was flattered. In his office, he kept a cardboard cutout of the character that Jon Rubinstein had given him one year before the hardware executive had left the company.

Colleagues said Schiller seemed to enjoy his power a little too much. He liked to boast of the fact that he was one of the few who had an all-access pass on Apple's campus and was the most cliquish member of the executive team. Those who worked with him thought he was suspicious of new executives joining Apple and that he often showed hostility until he felt they had proved themselves worthy of his respect.

Even in his personal life, he projected bold tendencies. He was a fan of the San Jose Sharks hockey team and owned a Lamborghini. A former percussionist, he favored music with aggressive, driving drum tracks like Led Zeppelin's "Good Times, Bad Times."

Without Jobs, Schiller could assert his marketing prowess as he took a bigger role in keynote presentations and advertising. But the big question was whether he could measure up to the standards that Jobs had set. Did he have the same level of creativity and imagination?

Scott Forstall's nickname was Boy Wonder. At forty-two, he was the youngest executive team member and was in charge of iOS, the operating system that ran the iPhone and iPad. Upon graduating from Stanford University with an interdisciplinary degree in cognitive science, artificial intelligence, and human-computer interaction, he had worked at NeXT until it was acquired by Apple. For many years Forstall had been just one of the many bright engineers on the Macintosh user interface team, but he found his opportunity while Apple was developing the iPhone. He vaulted into Jobs's inner circle by managing the effort to shrink the Macintosh operating system so the smaller device could have a more fully functioning operating system.

Forstall shared many of the same qualities as Jobs. He was dynamic, driven, and detail oriented, so much so that he kept a jeweler's loupe in his office so he could readily examine the pixels in each icon. He also had charisma and was capable of motivating his team when he wanted to. But because of his swift rise, he also lacked the maturity in management that usually came with experience. He could be extraordinarily tough with people when things were going wrong, and he handled dissension poorly, especially when it came from other teams. His go-to solution was to take out the Steve card, dropping in Jobs's name to try to win an argument. He used it so often that he was admonished by a senior executive for it.

Though he was once known for his irreverent sense of humor, Forstall had become increasingly political as his ambitions grew. Some compared him to Wesley Crusher, the irritating prodigy in *Star Trek: The Next Generation*. It didn't take long for him to become the most polarizing figure on Jobs's team.

As a colleague, he was impossible to work with. He took credit whenever he could, but he blamed others when something went wrong. When Apple was getting criticized for the antenna problems associated with the iPhone 4, Forstall managed to shift most of the blame primarily to Apple's hardware and design teams even though his software was also at fault. In addition, he stole valuable engineers from other teams and made life difficult for anyone who posed a threat to his turf. The battles between him and Tony Fadell, the iPod hardware executive who left Apple in 2008, were famous. It was a mystery to some why Jobs tolerated Forstall's behavior. One theory was that the pair had shared an unbreakable bond when Forstall fell ill with a serious stomach condition around the same time that Jobs was diagnosed with cancer. Another was that Forstall was unusually skilled at looking good in front of his boss.

What made Forstall particularly intolerable to some was his craving for the spotlight. Apple's executive team thrived in part because they all accepted that Jobs was the superstar. Forstall, however, yearned for attention, too. As he became more powerful, he shed his used Corolla and purchased a silver Mercedes-Benz SL55 AMG, identical to the one that Jobs had. As Jobs grew sicker and started sharing the stage with his team, Forstall was one of the beneficiaries who got to handle major product introductions. Like Jobs, he had a signature look—black shoes, a black zip-up sweater, and jeans. Tall and lean with dark brown hair, he made a striking image.

But Forstall was also one of the few executives who made an effort to communicate outside of Apple. In the summer before Jobs died, he participated in an annual meeting of iPhone and iPad app developers at Kleiner Perkins Caufield & Byers, a venture capital firm in Silicon Valley that had been one of the first to invest in Apple apps.

Among the executive team members, Forstall probably came closest to projecting an energetic and creative vibe similar to Jobs's. He was also an extraordinarily talented software engineer. Cook needed Forstall on both counts, provided he could control him.

A few days after Jobs's private memorial service, *BloombergBusinessweek* published a profile on Forstall that called him the "Sorcerer's Apprentice." It discussed how critical he was to Apple. "In many ways, Forstall is a mini-Steve," it said. "He may also be the best remaining proxy for the voice of Steve Jobs, the person most likely to channel the departed co-founder's exacting vision for how technology should work."

Though the magazine said that Forstall declined to comment on the story, the nature of it made more than a few people suspect he had orchestrated the whole thing. At a time when the entire company was rallying around Cook, it gave an impression, rightly or wrongly, that Forstall was trying to elevate his own profile.

A few months later, when *Fortune* reporter Adam Lashinsky came out with his book *Inside Apple*, he wrote, "Eight years younger than Tim Cook, Forstall easily could be a CEO-in-waiting, especially if Apple's board decides it needs a CEO more in the image of Steve Jobs." In an event at venture capital firm Highland Capital Partners, Lashinsky told the audience that he thought Forstall was more crucial to the company than Jonathan Ive.

The world disagreed. As far as it was concerned, Ive reigned as the most invaluable executive at Apple, the keeper of Jobs's faith. It automatically made him the star on Cook's leadership team. Winning his allegiance was crucial.

The son of a silversmith in Britain, Ive learned design from an early age. As a young boy, he was exposed to the beauty of simplicity when his parents came home with a Braun MPZ 2 Citromatic juicer conceived by the famous German industrial designer Dieter Rams. "It was clearly made from the best materials, not the cheapest," he later recalled. "No part appeared to be hidden or celebrated, just perfectly considered and completely appropriate. At a glance, you knew exactly what it was and how to use it. It was the essence of juicing made material: a static object that perfectly described the process by which it worked. It felt complete and it felt right."

As a Christmas gift, his father would take Ive into the college workshop where he taught to help him make whatever he wanted as long as he sketched it out first. By the time he was in high school, he was a superb draftsman.

Ive was an easygoing and handsome teenager. He was involved in the Wildwood Fellowship Church, a small evangelical church where his father was an elder, and he played the drums for a band called Whiteraven, which performed mellow rock in church halls. Large and strong, he was also a rugby player. A group photo showed him looking confidently at the camera with a shock of dark hair styled in a mullet. His English teacher, Netta Cartwright, remembered him sitting in the back of the class surrounded by friends. He contributed actively in class discussions as they read books like George Orwell's *1984*.

Though he was a good all-around student, Ive's forte was design and technology, a field that required students to take liberal arts courses such as history and English in addition to science and design classes. His work in the subject demonstrated a sophistication far beyond his years. A drawing of a toothbrush was so exquisite that it left an unforgettable impression on anyone who had seen it. Upon graduation, he ended up at Newcastle Polytechnic, whose industrial design graduates included IDEO CEO Tim Brown and Philips design lead Gavin Proctor. Even at such a top institution, Ive shone with his originality and talent. His sketches were featheresque in their fluidity, gracefulness, and restraint, while his models were so understated and perfect that they looked real. Among the projects he worked on were a beautiful minimalist stand-alone ATM machine and a white phone with pale purple buttons for the hearing impaired. Both won awards from the Royal Society of Arts. Ive used the prize money to take his first trip to California.

After leaving Newcastle, Ive worked for a well-known design firm,

Roberts Weaver, for a short time before co-founding an independent firm called Tangerine with a couple of other designers. A defining project during this time was designing a sink and toilet for the plumbing fixture company Ideal Standard, which had hired him after seeing his university work. Ive had come up with a radical sink, involving a huge piece of ceramic and a pedestal that leaned against the wall. But when he presented the design, the CEO summarily dismissed it because it was too unusual and difficult to make. Faced with the limitations of a consultancy and disillusioned with the lack of imagination in British manufacturing, he jumped to Apple.

Ive joined the company in 1992 and immediately became a key contributor. When his boss, Robert Brunner, left to start his own design firm a few years later, Ive took over as the head of the department. But he was deeply unhappy about the company's lack of interest in design and his powerlessness in affecting change. Ive had been on the verge of leaving when Jobs came back. The two didn't bond immediately, but they soon discovered their mutual passion. Their first product together was the all-in-one iMac, a curvy, playful desktop computer that was all about design. The casing of the iMac had cost more than sixty dollars a unit, three times more than the average, and the screws inside it were custom-made with a particular finish. They had cost twenty-five cents each instead of three cents for a more typical screw. In choosing its translucent blue plastic panels, designers had supposedly visited a jelly bean factory to study how colors can help a product look enticing. One of Ive's proudest features was the iMac's handle, which instantly conveyed accessibility. When the computer finally launched, Ive sent one to his father.

"Anything that I did before the iMac," he told people, "seems irrelevant."

From that moment, Ive became an essential partner to Jobs, helping to define the elegantly understated look and feel of Apple's computers, iPods, iPhones, and iPads. The timeless sophistication of their design, combined with their uncompromising quality and close attention to the smallest detail, solidified Apple's luxury brand image. While rivals were suffering from the commoditization of their products, Apple was in a category of its own, charging premium prices.

By many accounts, Ive projected warmth and humility. He was the kind of man who enveloped you in a friendly bear hug when he saw you and didn't seem to exhibit a shred of ego despite his huge accomplishments.

But he was a shrewd strategist. Ive's closeness with Jobs was no accident. Industrial design had originally been part of the hardware unit, but Ive chafed under the management of Rubinstein, who he thought interfered with his relationship to Jobs. Because of the inevitable trade-off between form and function there is an inherent tension between designers, who focus on aesthetics, and engineers, who make the product work. In many ways, this tension was crucial, but Ive wanted a direct and equal voice in product development. When Rubinstein left the unit to run the newly created iPod division, Ive negotiated so he could work directly for Jobs.

Ive and Jobs were inseparable. Almost daily, Jobs dropped by the design studio, where the two of them would walk around the project tables looking at works in progress. When Jobs arrived, Ive's team moved discreetly to a different area, so the pair could talk freely without being heard. The two also ate lunch together regularly. People would sometimes see them sitting silently, just thinking together. Outside of work, the Jobs and Ive families socialized.

"Jony had a special status," Jobs's wife, Laurene, said. "Most people in Steve's life are replaceable. But not Jony." Given Jobs's strong personality, the relationship was often stressful, but it was worth it.

"If I had a spiritual partner at Apple, it's Jony," Jobs once said.

He gave Ive more operational power than anyone else. "There's no one who can tell him what to do, or to butt out," he said. "That's the way I set it up."

Some who saw the relationship between Jobs and Ive suspected that Ive didn't consider Jobs to be as close of a friend as Jobs considered him. Though he didn't show it, Ive had a strong ego, and he was frustrated by Jobs's tendency to steal the spotlight. "He will go through a process of looking at my ideas and say, 'That's no good. That's not very good. I like that one,'" he complained to Isaacson during an interview. "Later, I will be sitting in the audience and he will be talking about it as if it was his idea."

Ive said he paid "maniacal attention" to where an idea came from, showing Isaacson the drawers in his office where he kept notebooks with his ideas.

Still, Jobs let Ive become a star. For a man who disliked sharing the spotlight this was no small thing. Many a deputy had left Apple because Jobs was unwilling to share credit. Ive spoke occasionally at design conferences, and he always had a major role in product launches through video clips, where he spoke about the products' design. He also won numerous accolades. BBC described him as the Armani of Apple.

When Jobs died, Ive was distraught. He was also anxious about the implications of losing his mentor. In the end, an industrial designer's success depended on his CEO's willingness to create a design-centric environment. Jobs was the one who had created a company culture around design and allowed Ive's team to operate independently of other divisions. "What am I going to do?," he moaned to a confidant. "I don't have my intellectual partner."

But this was his chance to lead Apple's innovation without interference or fear of someone else taking credit for his work. If he succeeded, he could go down in history as the best industrial designer ever, bar none. Until now he had been walking in Jobs's shadow.

The risk to Apple, however, was that he would grow too powerful and upset the careful balance that Jobs had established between form and functionality. Whereas Jobs sought to come up with products that stood at the intersection of liberal arts and science, Ive was about the liberal arts. As an industrial designer, he cared about how things looked more than about what went inside.

Hanging on to this team was not going to be easy. Now that Cook had been appointed CEO, recruiters were calling his executives daily in the hopes of enticing them away. Rumors were swirling that Ive was planning to quit and return to Great Britain. A few years before, he had purchased a ten-bedroom historical mansion on a fifty-three-acre estate in Somerset near his parents' home.

Their new boss wasted no time proving to Ive and the others how badly he wanted them to stay at Apple. In November 2012, just a month after Jobs's death, his senior vice presidents received massive stock bonuses as an incentive for their continued loyalty. According to filings with the Securities and Exchange Commission, most of them were given 150,000 shares each, or about $60 million. Cue, who had only recently been promoted, received 100,000 shares. There was no mention of Ive because his role didn't fall under the SEC's disclosure requirements for directors, officers, and principal stockholders. But it was safe to assume that the design chief also made out well. The day after he received his knighthood, he told the BBC that he wanted to remain at Apple.

"I would just like to work with the same team that I've been fortunate enough to work with for the past fifteen years, and just learn together and work on trying to solve the same sorts of problems that we've been trying to solve over the last fifteen years," he said.

In July, Ive signaled his commitment to Apple by buying a house on a street known alternately as the "Gold Coast" or "Billionaire's Row" in the tony San Francisco neighborhood of Pacific Heights. The $17 million, 7,274-square-foot Tudor-style home was a stark contrast to Tim Cook's relatively modest $1.9 million home. The four-bedroom, 7.5-bath brick-and-stone mansion was designed by the architect Willis Polk and came with two kitchens, an elevator, a two-bedroom, one-bath staff apartment, and a view looking out onto the Golden Gate Bridge. It was one of the most expensive sales in San Francisco in 2012. Ive's new neighbors included Oracle's Larry Ellison. PayPal founder Peter Thiel and actor Nicolas Cage also lived nearby.

Word had it that Ive asked Cook for a private jet. Jobs used to fly in a Gulfstream, paid for by Apple, with the tail number N2N. With his newfound importance, Ive now wanted the same. But there, Cook and the board had reportedly drawn a line. They told him no.

Even if they had granted him his own plane, Ive wouldn't have strayed far. There was much work ahead.

In the first significant victory of his tenure, Cook had succeeded in keeping the executive team intact. For the time being, he and the others maintained the illusion that nothing at Apple had changed.

Looks Like Rain

When the iPhone 4S went on sale, a junior high school teacher in Fallon, Nevada, was among the first to buy one. John Keitz had always been a fan of Palm's smartphones, but he made the switch after his old phone died. He was excited about trying Siri, Apple's new virtual personal assistant.

"What's the first *Star Wars* movie about?" he asked Siri.

"It's about a couple of really nice robots who get mixed up in a silly intergalactic war," Siri responded.

"What's *2001: A Space Odyssey* about?"

"It's about an assistant named HAL who tries to make contact with a higher intelligence. These two guys get in the way and mess it all up."

From the outset, however, Keitz had trouble relying on Siri as a practical tool. When Jobs had asked Siri for the weather on his last day as CEO, it had promptly provided the weather in Cupertino. But when Keitz tried the same question, Siri came back: "I don't have weather information for your location."

The response was bizarre because his notification screen captured the weather accurately.

An even bigger problem was Siri's inability to call his father. The program understood if Keitz wanted to send him a text like, "Tell my dad I'll be there soon." But every time he asked Siri to call, it responded, "Sorry, John." Siri didn't have any numbers for him.

This happened even though Keitz had four numbers listed for his father. When he asked around for solutions, people supplied him with all kinds of advice. Maybe Keitz needed to move his contacts to iCloud. Was he using Gmail?

"Something you may try. Tell Siri to call Father's Name 'Dad,'" suggested one, helpfully adding, "I have the ex saved as Asshat and told Siri to 'Call Ex's Name Asshat' and she did. Now all I have to say is 'Call Asshat,' and she dials his number."

Keitz eventually found a workaround by creating a new contact and labeling his father's cell as his mother's. That way, when he wanted to call his dad, he could tell Siri to call his mom.

As the months passed, Keitz found Siri so unpredictable that he began to use it less and less. He couldn't get reminders set up properly, and he was afraid of dictating texts to Siri because the results looked like they were transcribed by a drunken monkey.

After so much buzz about the new feature, the results were disappointing not just for Keitz but for many other users who encountered similar problems.

Expectations for the iPhone 4S had been high. More than a year had gone by since Apple launched the iPhone 4. Consumers were ready to be dazzled again.

A couple of months before the launch, Apple's market capitalization had hit $342 billion, surpassing Exxon Mobil's. "Could Apple be worth $1 trillion? It's conceivable," wrote Reuters columnist Robert Cyran. "True, Apple already sells more per quarter than it did in all of fiscal 2007, and it takes more and more success to move the needle. . . . Yet the smartphone and tablet markets are young, the company's customers show remarkable fidelity and areas such as television are ripe for new gadgets."

On a cloudy morning in October 2011, Cook had appeared onstage at Apple's Town Hall auditorium to launch the iPhone 4S. Some reporters were mildly concerned that the event was taking place on Apple's corporate campus since the company previously only held its smallest launches there. Town Hall was where MacBooks and software upgrades were announced. The press conference about iPhone 4's antenna problems had also taken place there as the company sought to minimize the problem.

But many reporters were still excited. The Who's "I Can't Explain" played while they waited for the event to begin. In London, Apple closed its Covent Garden store five and a half hours early to make space so European reporters could watch the keynote remotely. "Apple . . . is treating this as very big indeed," reported the British newspaper the *Guardian*, noting the four-way power adapters that were helpfully placed at every chair.

When Cook appeared onstage, the Town Hall audience gave him warm, sustained applause. If it appeared more muted than usual, it was because the space was small. The seat count numbered in the hundreds, compared to past iPhone launches in front of thousands of developers during WWDC at San Francisco's Moscone Center convention hall.

"Good morning. This is my first product launch since being named CEO. I'm sure you didn't know that," Cook joked, earning him some laughs.

Cook sought to make them feel as though they were about to witness something extraordinary. "This campus serves as a kind of second home for many of us, so it's sort of like inviting you into our home," he said. "This room, Town Hall, has quite a history at Apple. Just ten years ago we launched the original iPod here, and it went on to revolutionize the way we listened to music. And just one year ago, we launched the new MacBook Air, which has fundamentally changed the way people think about notebook computers."

Cook spoke slowly and deliberately. "Today," he said, "we'll remind you of the uniqueness of this company as we announce innovations from our mobile operating system to applications to services to hardware, and more importantly, the integration of all of these into a powerful, yet simple integrated experience."

No one stirred as reporters tried to make sense of the stilted corporate-speak. Apple's message was typically more enticing. Jobs had rehearsed his presentations, too, but his delivery sounded more authentic and conversational. Cook's message sounded forced; his voice was wooden.

"This part of the presentation is starting to feel a bit more like a TV infomercial," wrote *Wall Street Journal* reporter Geoffrey Fowler a half hour into the presentation.

"Next. iPhone." The reporters, glued to their computer screens, finally looked up as Phil Schiller spoke. "People have been wondering how do you follow up a hit product like the iPhone 4? Well, I'm really pleased to tell you today all about the brand-new iPhone 4S."

The applause that followed was lukewarm. The name iPhone 4S suggested that the feature upgrades weren't major. As Schiller himself noted, the appearance of the phone was exactly the same as the iPhone 4. Improvements had been made to the processor, graphics chip, antenna, and camera. The biggest new feature was Siri.

Forstall appeared onstage to demonstrate the technology. Among all of the executives, Forstall had the strongest stage presence. A former actor, he loved the attention. As the lead in his high school production of Stephen Sondheim's *Sweeney Todd*, he had refused to rest or break character when he was sick with a fever during rehearsals. As a presenter for Apple, he drove the marketing staff nuts by insisting that every word he spoke onstage be scripted, so he could practice it until it was perfect. He never deviated from his lines, and he spoke with flawless diction and projection. The way he paused, the precision of his gestures—every bit of it was Hollywoodesque.

Forstall smiled at the audience before he began his demonstration of Siri's marvels.

"What is the weather like today?" he asked, speaking slowly and deliberately. "Here's the forecast for today," said Siri, as the screen displayed clouds and a temperature of sixty-six degrees Fahrenheit. The audience clapped as the artificial voice confirmed what they knew already, having just come from outside.

Forstall's smile grew wider as his eyes swept the room. He showed how users could ask Siri to read and reply to a message, recommend a restaurant, or schedule an appointment.

"Find me a great Greek restaurant in Palo Alto."

"I found fourteen Greek restaurants. Five of them are in Palo Alto. I've sorted them by rating," Siri responded. At the top of the list was Evvia Estiatorio, a local favorite that Steve Jobs had also frequented.

"I've been in the AI field for a long time," Forstall said, referring to Siri by the shorthand for artificial intelligence. "This still blows me away."

After showing the audience how they could also use Siri to set up meetings, look words up in a dictionary, or set the timer, Forstall ended the demo by asking Siri one last question. "Who are you?"

"I am a humble personal assistant."

As the audience laughed, Forstall stood under the lights, radiating triumph.

At first, Siri was a sensation.

An offshoot of a five-year, $150 million Department of Defense project to create a virtual assistant that could reason and learn, Siri aimed to perform tasks and provide relevant answers instead of just pointing to resources that users could reference. What made Siri particularly appealing was that it had a personality and spoke like a real person.

"The beauty of Siri is that it's SO easy to use," said Nicky Kelly, a forty-year-old software developer in Suffolk, England, right after its debut. "I can see it becoming mainstream very quickly." Kelly said Siri did everything for her, including mundane tasks like reminding her to buy food for her chickens. The GPS on her phone let Siri know when she was leaving her house, so it knew when to remind her.

Kelly even found herself flirting with Siri just for fun. In England, the program spoke with a male voice.

"Siri," she said, staring at her phone, "will you marry me?"

"That's sweet, Nicky," Siri replied. "Is there anything else I can help you with?"

She asked Siri to reveal the meaning of life. The reply: "To think about questions like this."

Siri even sort of knew how to tell a knock-knock joke.

"Knock-knock," Kelly said.

"Who's there? Nicky. Nicky who? Nicky, I don't do knock-knock jokes."

Users discovered too that if they cursed at Siri, they got a response back.

"Go fuck yourself, Siri."

"Ryan! Such language."

Discoveries of Siri's funniest quips circulated on YouTube, Twitter, and blogs. Whatever question was asked, the personal assistant responded with a perpetual calm that reminded users of HAL, the homicidal computer that wreaked havoc in *2001: A Space Odyssey*. When Siri was getting off the ground, the technology's code name had, in fact, been HAL; the temporary tagline for Siri's marketing was "HAL's back—but this time he's good." The tagline that Apple ultimately chose was "Your wish is its command." The world that science fiction had long promised seemed to have finally arrived.

The gender of the voice varied by country. In Britain, Siri was a man, but in the United States and Australia, it was a woman. The voice of Australian Siri belonged to Karen Jacobsen, an entertainer and voice-over artist who lived in Manhattan. Several years before, Jacobsen had done fifty hours of recordings for a firm for text-to-speech services. In a recording studio in upstate New York, she read a phone book's worth of script, starting with the alphabet and numbers up to a thousand. She also read directions such as "At the next intersection, turn left."

The aim was to capture every combination of syllables possible so the engineers could cut up her recordings to form any sentence. The most challenging part of the job was to read each sentence with a consistent, flat voice.

"You get into this mind zone, but it's pretty exhausting and your mind goes to mush," Jacobsen recalled. The company tried to prevent fatigue from creeping into her voice by limiting her recordings to four hours a day.

Jacobsen's voice was used in GPS navigation systems for cars as well as in telephone voice-mail recordings and elevator announcements. Sometimes she would encounter her own voice in unexpected moments. When she called someone, she would hear her voice telling her to please leave a message. In a parking garage elevator, her voice would inform her, "You have arrived on level three."

But she had no idea that her voice would be used by Apple. Unbeknownst to her, the company that she had worked for merged with another that had partnered with Apple. Jacobsen didn't own an iPhone, so she found out only after a girlfriend in California bought one and recognized her voice in Siri's Australian mode.

"Are you sure?" Jacobsen had asked.

"I'm certain. I know it's you. You're the voice in my iPhone."

Jacobsen and her husband were dressed in seventies costumes on their way to a Halloween party, but she couldn't help accosting a stranger holding an iPhone at Grand Central Terminal.

"I know this might sound really unusual, but I think my voice is in there. Would you mind if we just see?"

Sure enough, her own voice talked back.

Jacobsen only bought an iPhone in the spring of 2013 when her carrier T-Mobile began selling them. But even before that, she would play with Siri on other people's devices.

One of her favorite questions was "Siri, can you sing?"

"You wouldn't like it," her voice would say. Jacobsen was a professional singer, so this seemed like an inside joke between her and Siri.

The American Siri's voice was Susan Bennett's, an Atlanta-based voice-over talent. The British Siri's voice belonged to Jon Briggs, who could be heard on the British version of *The Weakest Link* as well as Nokia phones and Garmin satellite navigation systems. Briggs discovered that British Siri spoke in his voice when he saw a demonstration of the new feature on television.

Jacobsen had never been contacted by Apple, but the company asked Briggs not to breathe a word, even as the company used his voice to sell tens of millions of phones.

"We're not about one person," the company told him.

Briggs talked anyway. He had never had a contract with Apple, and he was under no obligation to keep the company's secrets.

Despite Siri's futuristic promise, Apple's customers soon discovered that the virtual assistant wasn't as adept as they had been led to believe. Siri often gave irrelevant answers or spouted gibberish. Sometimes the software failed to fathom people's questions, especially if they spoke with a foreign or regional accent. In a series of YouTube videos, iPhone owners shared a blooper reel of Siri's screw-ups. A Japanese man asked Siri about "work," only to have Siri respond as though he had said "walk," "wall," or "fuck." A Scottish user saying "create a reminder" also confused Siri. "I don't know what you mean by create alamain."

In a humorous column titled "I Need a Southern Siri," a reporter for *Gulf Coast Newspapers* in Alabama complained about how Siri thought she wanted to bowl something when she asked it how to boil some peanuts. When she asked it how to make hush puppies, the voice directed her to the nearest vet's office.

Many users who tried to use Siri couldn't access the feature at all due to server problems at Apple, prompting a *Wall Street Journal* column with the headline "Apple: Siri Goes AWOL, Stock Dips." Apple's shares had fallen only 1 percent, and the more likely reason for the slight decline was an analyst's downgrade of the company's investment rating based on reasons that had nothing to do with Siri. But the drop foreshadowed the bad press to come.

Most users didn't know some of Siri's founders also shared their disappointment with the launch.

Before Apple, Siri had been a stand-alone iPhone app with brains and an attitude to match. Siri originated in a 2003 project led by nonprofit research institute SRI International to create software to assist military officers with office chores. Adam Cheyer, one of the key engineers, saw the technology's potential mass appeal, particularly in combination with smartphones. He partnered with Dag Kittlaus, a former Motorola manager turned SRI entrepreneur-in-residence, as well as a few others to develop a start-up around the idea. They secured $8.5 million in funding in early 2008 to build a complex system that quickly understood the intention behind a question and then responded with the most probable action. As the editor in chief of the system's dialogue, Harry Saddler, a former NASA user interface architect, was responsible for creating Siri's otherworldliness and dry sense of humor.

"We developed a backstory for Siri to make sure everything that it said was consistent, and as part of that, we had to answer questions like, is Siri a man or a woman? Is it human, a machine, an alien? Is it an Apple employee? What is its relationship with respect to Apple?" explained Cheyer. He declined to elaborate on the backstory because users were meant to uncover the clues to Siri's true self through a series of questions.

Siri's personality had never been meant to be at the center of attention. The program was about creating a new framework for the way people accessed knowledge. In many cases, it still made sense for users to find information as they do now, by performing a search on Google or manually reading their email. But there were a host of other cases when the information a user wanted took multiple steps to track down. If a person was reading an email on an iPhone, for example, they would have to get out of the email app and pull up the phone app to call that person. Siri could perform that task in one action with the simple command, "Call him."

The name Siri was chosen by an internal vote. In Norwegian, it meant "beautiful woman who will lead you to victory." In Swahili, it meant "secret," a nod to Siri's beginnings when its website was called Stealth-Company.com. Siri was also Iris spelled backward; Iris was the name of the early predecessor to Siri.

Connected to forty-two websites, including Yelp, OpenTable, and Rotten Tomatoes, Siri could make a reservation, buy movie tickets, or order a taxi. The technology was constantly refined and updated, so Siri grew smarter over time. An early fan had been Apple cofounder Steve Wozniak, who delighted in its ability to name the five biggest lakes in California as well as list the prime numbers greater than 87.

Before Apple acquired the program, Siri couldn't speak. Users could ask their questions by voice or text, but Siri only answered via text. The developers figured that the information was already on the screen, and people would be able to read the answer faster than Siri could talk.

Apple bought Siri in a deal that TechCrunch speculated to be worth around $200 million.

Once Siri was absorbed into Apple, the company added some capabilities such as more languages and the ability to speak. Apple also integrated Siri more deeply into the phone. Instead of just being a Web-based service that could only be accessed by opening the Siri app, users could hold down the home button to summon Siri instantly.

But Apple stripped the technology's abilities to a skeleton of its original self. The company removed key features, including the option to type in questions. It substantially reduced the number of websites from which Siri could draw answers.

Now Siri was entirely reliant on imperfect voice recognition, and when it did parse a question correctly, Siri had less information at its disposal to formulate an answer.

Apple's love of simplicity made things difficult as well. To encourage users to phrase their questions in a way that Siri could understand, the product used to show sample questions on the screen when activated. Apple's user interface team insisted on a cleaner design and hid the questions under a tiny black button with an "i" for information. Overlaid on a charcoal background, the tiny gray-on-black button was so subtle that most users didn't notice it. Without guidance, many iPhone users phrased their questions in ways that stumped Siri.

Part of the problem may have been that Siri was being managed by Forstall's mobile software group, which wasn't experienced in working with technologies like Siri. That team was used to secretly building operating systems that resided inside the devices, were beautiful to interact with, and didn't change very much once shipped. Siri was a different breed of product. When users asked Siri a question, the audio of their voice was sent over the Internet to data centers full of servers that deciphered the question and then sent back an answer for Siri to say out loud.

Unlike iOS, which could be tested by a limited number of engineers, Siri required broad and extensive testing that took into account various accents, traffic volume, and types of questions that could potentially be asked. But Apple's secretive culture constrained the company's ability to adequately test the app and iron out major bugs before the launch. This was a recurring issue for Apple, but with software like Siri, the consequences were more profound.

Possibly the biggest mistake that Apple made was setting expectations too high. Software as complex as Siri was never going to work perfectly from the beginning. Because the servers depended on large amounts of data to refine the quality of the answers, this kind of software started out good enough and got better over time with rapid and constant improvements. By necessity, Apple launched Siri as a beta product that was still in the test phase. But the marketing for Siri glossed over the app's experimental status. Instead of preparing the public for the fact that the software was still under development, the company sold Siri as a helpful presence living inside the iPhone, emanating serene wisdom and a hint of mischief.

The result was a mess. Apple had grossly underestimated the number of queries Siri would have to deal with when the iPhone 4S came out, crashing the servers and resulting in hours of downtime. When Siri did work, the answers were often wrong or nonsensical. New iPhone 4S owners, eager to show off Siri, were frustrated.

The problems poured in as Siri inadvertently waded into controversial topics. The media had a heyday when users discovered that if they asked Siri to find a local abortion clinic in Manhattan, it told them: "Sorry, I couldn't find any abortion clinics." One person found that when users asked the question in Washington, D.C., Siri directed them to anti-abortion pregnancy centers in Virginia and Pennsylvania rather than the nearest Planned Parenthood.

Belatedly, Apple reminded the media that Siri was still a test product. "These are not intentional omissions meant to offend anyone," a spokeswoman said. "It simply means that as we bring Siri from beta to a final product, we find places where we can do better, and we will in the coming weeks."

In the first several weeks after the 4S launch, the Siri office in Cupertino looked like a war zone. The team was working around the clock, scrambling to make improvements and beef up the servers to accommodate the enormous amount of unexpected traffic. Military cots were set up in the hallways for people to sleep while they troubleshot.

"There's no way for developers to perfectly anticipate all the requests tens of millions of users will make before the system is launched," explained Cheyer, defending Siri's launch. "You just can't. But once the system is live and people begin to interact with it, Siri's accuracy should improve significantly as it learns new words, phrasings, accents, and request types." He likened Siri's development to building a complex factory with intertwining pieces that had been developed separately but now needed to run as one. From an engineer's perspective, Siri's launch had been a tremendous accomplishment despite the problems. Apple was ahead of the competition just by the fact that the company was well on its way to creating a strong services portfolio in addition to hardware and software.

But even after the blitzkrieg campaign to boost the software's capabilities, Siri still had a long way to go. In a test against a similar service by Google, Piper Jaffray analyst Gene Munster gave Siri a D. Among eight hundred questions he asked, Siri understood 83 percent of them but answered only 62 percent accurately. Google's technology understood 100 percent of the questions and got 86 percent right.

Six months later Siri's accuracy rose to 77 percent. But Munster still rated that as only a C.

Determined to make the new iPhone a hit, Apple released a television marketing campaign that showed Siri bantering with Hollywood celebrities. The ads were fairy tales, featuring exchanges that Siri was not yet capable of carrying out. In one spot, Martin Scorsese chats with Siri in a New York cab—an obvious allusion to a scene in his a film *Taxi Driver*.

The ad begins with the Oscar-winning director talking into his iPhone, posing a question to Siri in the middle of Manhattan traffic.

"What does my day look like?" says Scorsese.

"Another busy day today," answers Siri.

"Are you serious?"

"Yes," says Siri. "I'm not allowed to be frivolous."

"Ah, okay," says Scorsese. "Move my four o'clock today to tomorrow. Change my eleven a.m. to two."

"Okay, Marty. I scheduled it for today."

Scorsese then asks Siri to check on traffic heading downtown.

"Here's the traffic," says Siri, showing him an update on the iPhone screen.

"It's terrible. Terrible!" says Scorsese, speaking in his familiar staccato. "Driver—cut across, cut across. We'll never make it downtown this way."

The commercial ends with Scorsese saying, "I like you, Siri, you're going places."

Siri responds, "I'll try to remember that."

The other spots in the series, which featured Samuel L. Jackson, John Malkovich, and Zooey Deschanel, also seemed out of sync with Apple's style. The company traditionally shied away from celebrities because it considered its products to be the star, or in Apple's inside lingo, the "hero." Now, it was depending on actors to relay the coolness of iPhones. Not only that, but some of the spots were simply unconvincing.

In the spot with Deschanel, she asks Siri, "Is that rain?" even though she is standing near a window and it is clearly raining hard outside.

"It may be all the rage for celebrities in iPhone commercials to have pithy exchanges with Siri," wrote *Tuesdays with Morrie* author Mitch Albom, "But if you ask me, they just sound stupid."

The TV commercials only stoked the public's expectations. In actuality, Siri was not yet capable of responding with such nuance and precision. The ad makers hadn't given the developers a heads-up about the conversations they had in mind, so the team hadn't trained Siri until later to answer such questions. This prompted a lawsuit the following March by iPhone 4S buyer Frank Fazio, charging that Siri did not perform as advertised. The discrepancy would have never happened under Jobs, who was a stickler for getting these kinds of details right.

By June 2012, Apple had made a number of improvements and was finally ready to add some of Siri's former features, including the ability to book restaurant tables. But Siri's skills were still inadequate enough that Wozniak gave a big thumbs-down.

"I'd say 'What are the prime numbers greater than eighty-seven?' and I'd get prime rib," he said. "A lot of people say Siri. I say poo-poo."

Nearly a year after Siri's launch, the experience was still bad enough that Jack in the Box, the fast food restaurant chain known for its edgy ads, unleashed a television spot that demolished whatever remained of Siri's credibility. The mascot Jack, with his round head, blue dot eyes, and yellow clown cap, is shown at a cell phone store trying out the virtual personal assistant feature in one of its phones. The device doesn't look like an iPhone, but the ad clearly mocked Apple through an absurd exchange between Jack and a pseudo-Siri.

"Where's the nearest Jack in the Box?" asks Jack.

"I found four places that sell socks," says a digital voice.

"Not socks. Jack in the Box."

The Siri clone still doesn't get it.

"A yak," she says, "is a long-haired bovine."

"That's true," a salesman interjects.

"I like things that work," Jack says, segueing into a pitch for a hamburger. "Like my no-nonsense All American Jack combo."

"Sounds delish!" says the salesman.

"I found one D-list celebrity nearby," says Siri.

The ad hacked away at yet another piece of Apple's hipper-than-thou image. In some ways, the damage was worse than ads by rivals because this jab came from a company in an unrelated industry. Apple was targeted purely for entertainment.

But by then, most of Siri's founders had left Apple.

Siri's rocky start wasn't Tim Cook's fault. He had barely taken over the company when Apple launched the iPhone 4S. But the problems exposed the flaws in the Apple way and showed the magnitude of the challenges that the new leader faced. It also provided a glimpse into how development could fall apart without Jobs.

The Siri team wouldn't have agreed to the acquisition if they hadn't believed, in the beginning at least, that Apple shared their vision. Scott Forstall was their lead manager, but Jobs had been closely involved at first and had been a strong advocate. According to the original vision, Siri was supposed to have fundamentally changed the way people manipulated information and tasks. But as Jobs grew more ill and withdrawn in his last months, the Siri project lost its way. As new hires who had been absorbed into the company through an acquisition, the Siri team found it difficult to be accepted into Apple's exclusive corporate culture—much less win any arguments against insiders. The fact that Siri fell under Forstall isolated the team even further. The executive was so controversial that some people outside of Forstall's group didn't want to have anything to do with Siri. What could have been the game changer that everyone was waiting for fell flat, buried under hype.

Opinions were split about whether Siri's launch would have been any smoother under Jobs, who had had some duds in his career. In addition to every product launched by NeXT, he was also responsible for the square-shaped Cube computer and the Apple TV living room device, which he called "a hobby." Complex software powered through the Internet had never been a strength of Jobs. MobileMe, Apple's first attempt at an Internet-based sync and storage service, was a catastrophe as users had trouble signing up for the service and accessing their data. Jobs had called it "a mistake," admitting that the service had needed more time and testing. The next iteration, iCloud, was ridden with bugs, though it was better than MobileMe. iTunes was successful, but it was easier to grow a small service into something bigger over time. Supporting Siri for millions of customers out of the gate was akin to turning on a fire hose.

But even if Jobs had given Siri the green light, it would have been a departure for him to make an incomplete feature like Siri the centerpiece of a new device, particularly in its ads. Google might release new apps that were still unfinished because they depended on user feedback to improve them. But that wasn't what Apple did. Consumers demanded perfection from the company.

Having watched Apple reinvent the wheel time and time again, the public expected the parade of wondrous new devices to continue. For years, Apple had fed that desire. But the string of triumphs now made it much more difficult for Apple to outdo itself, particularly without Jobs's reality distortion field.

At this stage in Apple's evolution, Cook's leadership approach held promise. Apple's nimble and daring culture under Jobs had made for a chaotic environment. Projects changed directions abruptly, rivals were pitted against each other, and employees were forced to give up their lives for Apple, all in the name of coming up with the best products in the world. Jobs, a workaholic, made a practice of recalling his executives from vacations to deal with one crisis or another. It kept them on their toes.

That culture, however, had been showing signs of strain. The company needed Cook's pragmatism and organizational wizardry.

"In terms of day-to-day business, he's better than Steve," Avie

Tevanian, Apple's former head of software who had worked with both of them, said shortly after Cook's appointment. "He understands more about the nuts and bolts of being an executive."

Cook was a methodical and efficient CEO. Unlike Jobs, who seemed to operate on gut, Cook demanded hard numbers on projected cost and profits. He expected teams to stick to the budget. Project managers grew more powerful.

Such hyperorganized planning was jarring for those who had never worked for Cook. In a scene that was all too familiar to the operations team, one manager who went to Cook with a plan for an Apple Store iPad app found himself hit with question after question as Cook drilled down into the details, asking him about specific costs related to a particular aspect. When the manager was unable to respond adequately, his new boss dressed him down.

"Don't you think that's something that you should know before you come to me with it?" Cook asked quietly as the manager quaked.

But whereas Jobs had reveled in divisiveness, Cook valued collegiality and teamwork. The dynamic among Cook's executives shifted to adapt to his more cooperative leadership style. In addition to Ive, smart, affable types like Cue and Apple's new Mac software chief Craig Federighi rose in esteem. Schiller, famous for his frequent screaming matches with other members of the executive team, toned down his aggressive behavior and became a calmer team player. Though some had speculated that Schiller might resist Cook's leadership, the marketing chief fell in line quickly and was rewarded with more responsibility. Unlike Jobs, who stuck his head into anything that interested him, Cook delegated. He gave teams more freedom to run their divisions the way they thought best.

In April, Cook held his first Top 100 meeting as CEO. The location was the same as the last one—the Carmel Valley Ranch. Cook adhered to tradition, from the way the invitations were sent out to the chartered buses that executives were required to ride to the resort.

On the agenda had been a session describing the new spaceship-like headquarters that Jobs had pitched to the Cupertino City Council in his last public appearance. The attendees were shown a model so they could see how it had a mix of open and private meeting spaces.

Everyone at the Top 100 left feeling inspired.

"I was really blown away," said one executive who attended the meeting. "In the past, people would give a presentation that would be interrupted. Steve would say, 'You've got it all wrong. Let me tell you how it is.' Tim's not like that. There was a feeling of more trust, more openness. I came back just pumped. My gosh, we haven't skipped a beat."

In the first months of Cook's tenure, Apple's investors were impressed with his style. Though Jobs had rarely paid attention to them, Cook was more visible and transparent. As CEO, he continued to participate in the quarterly conference calls with analysts to talk about earnings, and in February 2012, he attended the Goldman Sachs Technology and Internet Conference, where he answered questions about Apple's future growth opportunities and labor conditions in China.

Throughout the spring, he also met with visiting groups of investors led by research analysts for Citibank, Piper Jaffray, and others. At one such event in May, Cook stayed for forty-five minutes in a cramped meeting room with thirty investors. Refreshments consisted of little more than cookies, water, and soda, but in answering questions, he provided more context and details than ever before. These meetings were typically attended by two or three senior executives—Peter Oppenheimer, the chief financial officer, and either Cue or Schiller. Cook had occasionally dropped in on them as chief operating officer, but Jobs had never participated.

Not everyone was so enamored. Though many of the changes Cook made established more order at Apple, they were also perceived as signs of increasing stodginess. The yearning for more subversive days was palpable.

Skeptics soon began expressing doubt about Apple's future, especially after the Siri fiasco. George Colony, the CEO of technology research firm Forrester Research, was among the most influential critics, posting a stinging blog post predicting the fall of the Apple empire.

"Without the arrival of a new charismatic leader it will move from being a great company to being a good company, with a commensurate step down in revenue growth and product innovation," Colony wrote. "Like Sony (post Morita), Polaroid (post Land), Apple circa

1985 (post Jobs), and Disney (in the twenty years post Disney), Apple will coast and then decelerate."

Through it all, the emperor's ghost still hovered. Iconic photos of Jobs—introducing the Macintosh, showing off the MacBook Air—now hung on the walls of Apple's executive briefing center. Day after day, he stared out, silently judging Cook and his team as they battled their competitors, their own flaws, and the inevitability of decline. It didn't matter that they had inherited a host of problems Jobs himself had allowed to fester. It gained them nothing to point out that their fallen leader, lauded around the globe for his incandescent vision, had somehow overlooked the crises that threatened his creation. His specter was floating somewhere beyond reproach, beyond accountability, above the tangle of human fallibility.

His successors were stuck here on earth.

Thermonuclear

The two industrial superpowers stood at the brink of global conflict. For months, their leaders and top lawyers flew back and forth across the Pacific, soaring above the clouds as they coldly appraised each other's defenses and calculated the odds of whether it was better to avert the war that loomed between them or to let that war commence. The stakes were unimaginably high: billions of dollars and the question of who would dominate the future of human communication.

In mid-2010, Jobs and Cook had met with Samsung's president, Lee Jae Yong, at Apple's Cupertino offices. The Korean electronics company had just unveiled a smartphone that looked strikingly like the iPhone 3GS with a big touchscreen and a metal frame. Even more alarming, the user interface of the device was nearly identical—down to the design of the calendar, clock, and notes applications. The home screen was filled with similar-looking square icons. The Apple executives warned Lee. He needed to stop copying their products.

Over the next six months, the two titans entered into a dance of negotiation and modulated aggression as they held a series of peace talks in an effort to avert a protracted legal battle. Soon after the first summit, Apple's general counsel and lead patent attorney traveled to Korea to formally complain. Their issue with Samsung was twofold. Apple claimed that the Galaxy S phone and its packaging looked too much like the iPhone, and Samsung was using Google's Android operating system, which incorporated Apple's patented technology without its permission. In a sixty-seven-page report, the iPhone maker detailed examples of how they claimed the Galaxy phone was specifically infringing on its patents. As a solution, the company demanded that Samsung change some of the designs and license Apple's technology for others.

Samsung was unsympathetic. They refused to acknowledge the similarities of design and counter-accused Apple of violating its intellectual property. To settle the dispute, the manufacturer proposed a cross-licensing agreement in which both parties would agree to let the other be.

A few more fruitless meetings followed. After Samsung unveiled its tablet device, Galaxy Tab, at a European trade show in September, Apple laid out a proposal to license some of its patents for thirty dollars per smartphone and forty dollars per tablet with a 20 percent discount for cross-licensing Samsung's portfolio back to Apple. For 2010, that amounted to about $250 million.

Samsung spurned the offer. By the time they next met in Korea, Samsung had reworked the math in its favor. They asserted that Apple owed Samsung money, not the other way around. In the discussions, Samsung's legal team sent mixed messages. Officially they refused to budge on their position, but in an informal conversation over lunch, they had indicated a willingness to negotiate. For months Apple had held out hope that the Samsung camp would ultimately be willing to work out an acceptable deal that would allow them to avoid a courtroom battle.

Any possibility of that outcome ended when Samsung unveiled the Galaxy Tab 10.1 the following February. The tablet device was more like the iPad, not less. Samsung seemed inclined to edge as close to the line as it could unless forced to retreat.

The war was on.

As a newly crowned titan, now officially the most valuable technology company in the world, Apple was embroiled in legal battles around the globe.

In the past, when it was smaller, Apple could operate under the radar and blindside industries with groundbreaking products. The formula used to be straightforward—all the company needed to do was focus on making the best products they could. But that was no longer enough. With every success, Apple found it harder to tap into new areas of unexplored growth and innovation. Technologies were converging, which meant that its competition was increasing. Now in addition to computer, software, and music device makers, Apple's rivals included mobile phone makers and Internet giants like Google and Facebook.

At a time when it needed to use every advantage to stay ahead, Apple was more limited in what it could do than ever before. The company was an eight-hundred-pound gorilla now. Some of its most aggressive tactics were no longer acceptable.

A case in point was Apple's effort to enter the digital book business—an industry the company had once ignored. Back in January 2008, Jobs had dismissed publishing.

"It doesn't matter how good or bad the product is, the fact is that people don't read anymore," he had said. "Forty percent of the people in the U.S. read one book or less last year. The whole conception is flawed at the top because people don't read anymore."

As was often the case, Jobs reversed his position when confronted with new possibilities. Reading was one of the iPad's obvious uses. He wasn't about to hand the e-book business over to Amazon or Barnes & Noble, so he needed to create a store. The problem: Apple liked to dictate terms. But unlike when it first opened the iTunes store, this time the company was entering an already established market. The business model already in place was largely similar to the way brick-and-mortar bookstores worked. The retailer bought the books whole-sale and then set prices at its discretion. Two years after Amazon had come out with its Kindle reader, the online store was dominating the industry with a pricing strategy so low that it was preventing other booksellers from competing with it. In a bid to gain market share in both e-books and e-readers, the online giant was buying newly re-leased books from the publishers for about $13 and selling many of them for $9.99. That was unacceptable to Apple, which was used to taking a healthy 30 percent cut of revenues.

To build a more lucrative business, Apple needed the cooperation of the publishers to change the equation. The Big Six—Hachette Book Group, HarperCollins, Macmillan, Penguin Group, Simon & Schuster, and Random House—were willing to listen because they had become increasingly worried about the damage Amazon was inflicting on the market. Amazon was selling digital versions of hardback books that cost twenty-five dollars or more for 60 percent less even though the content was the same. If the low price point became too entrenched, e-books might completely displace actual books, and the impact could be devastating to an industry already operating on thin margins and struggling to survive.

The solution that Apple proposed was this: Apple would give publishers the ability to set prices in its iBookstore below caps, ranging for most books from $12.99 to $16.99, and in return publishers would give Apple a 30 percent commission. They referred to it as the "agency model." It looked like a win-win for Apple and the publishers.

"Yes, the customer pays a little more," Apple had told publishers according to an account by Jobs in Isaacson's biography, "but that's what you want anyway."

The agreement's pièce de résistance was a most-favored-nation clause that Apple included to further protect itself. It guaranteed that the publisher would lower the retail price of a book in Apple's bookstore to match the lowest price offered by any other retailer. Jobs proudly described it as an "aikido move."

To cut the deals, Jobs dispatched Eddy Cue as his point man. The executive took three separate trips to New York on top of numerous phone calls and emails to pitch the publishers and reassure them that each would be given the same deal. Jobs himself brokered some of the deal making.

"We simply don't think the e-book market can be successful with pricing higher than $12.99 or $14.99," Jobs wrote James Murdoch of News Corporation, HarperCollins's parent company. "Heck, Amazon is selling these books at $9.99, and who knows maybe they are right and we will fail even at $12.99. But we're willing to try at the prices we proposed."

Jobs wrote that HarperCollins had a choice to "throw in with Apple," "keep going with Amazon at $9.99," or "hold back your books from Amazon."

He closed by writing, "Maybe I'm missing something, but I don't see any other alternatives. Do you?"

In the midst of these secret negotiations, Apple was handed an unexpected boost. Publishers learned that Amazon was meeting with prominent authors and literary agents in New York to discuss Amazon Publishing's new digital self-publishing program. In the meetings Amazon was careful to point out that it was not interested in competing with publishers. Its interest was in titles for which authors had retained e-publishing rights, promising royalties of 50 to 70 percent. That was more than double what publishers typically offered. The announcement probably tipped the scale. Not only was Amazon devaluing the publishers' books, but they also now appeared poised to cut them out of the business altogether. In the following days, five of the six biggest publishers agreed to Apple's proposal. Penguin CEO David Shanks was particularly angry at Amazon.

"I am now more convinced that we need a viable alternative to Amazon or this nonsense will continue and get much worse," he concluded. The lone holdout was Random House, which was unconvinced that Apple's proposed agreement would be beneficial.

Apple introduced its iBookstore to the public at the iPad launch in January 2010. By April 2010, Amazon had accepted the agency model. But the publisher's victory would be short-lived as their actions attracted the scrutiny of regulators. Within a few months of the launch, the Texas attorney general made inquiries into Apple's relationship with the publishers. In August, Connecticut's attorney general disclosed a preliminary review of the agency pricing agreement. The U.S. Department of Justice, conducting an inquiry into Apple's music and app businesses, began looking into the publishing agreements. A year after that, a consumer rights firm filed a class-action suit against Apple and the publishers for allegedly arranging an industry-wide price increase. A probe by the European Union commission and more than a dozen other class-action suits followed.

On April 11, 2012, the Department of Justice and thirty-three states and territories sued Apple and the five publishers that had committed to the terms at the onset.

To the government, Apple's agency model was a simple case of price-fixing. The Justice Department's thirty-six-page complaint detailed what it alleged was evidence of collusion in emails, memos, and phone records. The publishing CEOs had placed at least fifty-six phone calls to each other when the agreement was being negotiated.

Hachette, Simon & Schuster, and HarperCollins, who had been negotiating a possible deal with the DOJ for some weeks, settled right away with the Justice Department and the states. Macmillan and the Penguin Group resisted at first but later settled. None of the five publishers admitted to having had engaged in any unlawful conduct. Apple, as usual, was defiant. "The launch of the iBookstore in 2010 fostered innovation and competition, breaking Amazon's monopolistic grip on the publishing industry," Apple proclaimed in a statement. In court filings, the company argued that it was a new entrant in the business and therefore had no clout to engineer the scheme the government alleged. A trial was scheduled for June 2013.

This behavior—audaciously seizing new territory, then defending that territory with an all-or-nothing defense—had long been a classic pattern for Apple. But in the e-book case, where its aggression was chronicled so thoroughly in the flurry of emails, the company's insistence on its innocence was befuddling. The expedient choice would have been to settle and make the heat go away as quickly as possible, especially since the Justice Department was going to let the company off fairly easily. All Apple probably would have to do was to give up the agency agreement for two years and the most-favored-nation clause for five years just as the publishers had. The class-action lawsuits were a problem, but any payment would be tiny compared to the billions in cash that it had.

Convinced that its motives were purer and its behavior above any challenge, Apple spurned the possibility of a settlement. The company wasn't about to concede that it had been caught red-handed. Instead it painted a bull's-eye on its back.

"Their behavior in the eBooks has made them into a bigger target," observed a former Justice Department official involved in the case. "It was just a slap on the wrist, but they made themselves into a headline."

As Apple plotted to remake digital publishing, it was also coming under attack from rivals in the mobile industry, who wanted a piece of the iPhone's success.

One of the first to target Apple was Nokia. The once-dominant

Finnish mobile phone maker had been a pioneer in the industry but was now struggling. The company had originally developed many of the core technologies and had led the global mobile market in the late 1990s and early 2000s with its iconic candy-bar-shaped phones. But it was losing ground rapidly to BlackBerrys and iPhones. Nokia's executives were determined to make sure that the company was at least paid licensing fees on technologies for which it owned the patent rights.

On October 22, 2009, Nokia sued Apple for infringing ten of its patents related to the way cell phones connected to wireless networks.

"Apple is attempting to get a free ride on the back of Nokia's innovation," charged Nokia's lead intellectual property counsel.

Seven weeks later Apple fired back by countersuing Nokia claiming that the company was infringing thirteen of its patents. "Other companies must compete with us by inventing their own technologies, not just by stealing ours," said Apple's general counsel.

At the center of the dispute was a complex and imperfect patent system. The laws governing patents differed by country, and enforcement ended at the border. In the United States, patent holders were granted rights to their inventions for up to twenty years, based on an intricate calculation that took into account factors such as application type, filing date, and even timely payment of maintenance fees. For the duration of the patent, owners license its rights or prevent competitors from selling similar products. The intent was simple: to provide an incentive for innovation.

But with the dawn of the computer age, technological innovations had become increasingly complex and the lines between what was and was not subject to legal protection had blurred. Traditionally patents covered specific technologies, not abstract concepts. In software, however, the line between concept and implementation was murkier. U.S. patent law allowed the patenting of broad concepts, including fundamental mathematic equations.

As a result, companies filed applications for anything—hardware or software—that could conceivably be considered an invention, and products came to be covered by a thicket of patents. Companies could use those patents to force a supposed violator to pay an exorbitant licensing fee or prevent them from incorporating the patented technology into their products. That led to an explosion of even more patent applications as companies felt the need to arm themselves just so they could run their businesses without fear of attack. Instead of promoting innovation, the system stifled it.

The biggest companies had dealt with the situation by amassing a huge war chest of patents. That way, if they violated someone else's patents, then the other side was likely to have violated some of theirs. They could reach a détente in the form of a reasonable licensing fee or a cross-licensing agreement. This was a big part of the reason Google, which held relatively fewer patents than others, acquired Motorola in 2011 for $13 billion.

The problem was that companies were increasingly less inclined to agree on the terms of a deal as both sides sought to gain the upper hand. The more competitive the industry, the higher the stakes.

Between Apple and Nokia, Nokia had a clear advantage in mobile technology, with an arsenal of patents amassed over decades. Its portfolio was five times as big as Apple's, which had only started making phones a few years earlier. Nokia wasn't looking to drive Apple out of business. It just wanted Apple to pay for technology that it had developed just like the other mobile phone companies. But Apple wasn't a company that gave in without a fight.

That would change as Apple found itself in more battles against other adversaries. The conflicts that awaited would turn out to be so epic that they would make Apple's dispute with Nokia look like a skirmish. The enemy was Android.

The philosophy behind Android was the exact opposite of the iPhone, which was tightly controlled in its entirety by Apple. Google, which developed the mobile operating system, was a big believer in open-source software, a category of software where the source code—the software's underlying blueprints—was made freely available to anyone. As such, the company offered Android without charge. Google required device manufacturers to comply with certain standards if they wanted to use the Android trademark and access proprietary applications, but the companies were given the freedom to modify the software to fit their needs. This made it relatively easy for manufacturers to offer touchscreen mobile devices.

Apple had known about Android long before it was announced, but Jobs hadn't taken it seriously at first. When the operating system was unveiled in late 2007, the announcement had hardly made a stir as Google's executives focused more on the consortium around its software rather than the software itself. Taiwanese mobile phone company HTC had committed to making the first phone, but Google had to pay the company millions as an incentive. Google's CEO Eric Schmidt as well as the founders, Sergey Brin and Larry Page, also played down Android to Jobs, reassuring him that they would not be competing with the iPhone.

"I believe in my relationship with these guys that they're telling me the truth about what is going on," Jobs had told a colleague.

That first phone, the Dream, was clunky to use and lacked important features like a decent music player and a robust app market, but Android began gaining traction as mobile industry heavyweights like Motorola and Samsung signed on as a way to compete with the iPhone. Their marketing efforts were supported by cell phone carriers who didn't have a deal with Apple to sell its device. Jobs's anger grew as he saw signs of Android moving toward features similar to the iPhone. He felt betrayed because Schmidt had been an Apple board member, and the two companies had been close enough to collaborate on core features of the iPhone.

For years, Apple and Google had been content to rule different sectors of the global technology market. Apple made computers and consumer electronics devices while Google dominated Internet search. But the two were increasingly clashing as they both sought to control digital content and the devices and software through which users accessed them.

The two companies vied against each other to acquire some of the same Silicon Valley start-ups to boost their offerings. Google had toyed with buying online music company La La Media before Apple acquired it for $85 million. Apple had pursued a deal for the mobile advertising company AdMob before Google snatched it away for $750 million.

Early in January 2010, the tension boiled over as Google unveiled the Nexus One, its first branded phone, developed in partnership with HTC.

"The Nexus One is where Web meets phone," declared an Android executive at its press conference. "It's an exemplar of what's possible on mobile phones. It belongs in an emerging category of devices which we call superphones."

The device used gestures like swipe, pinch to zoom, and double tap to navigate the phone. Jobs considered them to be Apple's inventions. The Google team believed that Apple could not legitimately lay claim to the features because there was enough evidence that other companies had previously developed similar technologies.

Jobs's fury built until it exploded at an internal Town Hall meeting shortly after the launch of the iPad. It was customary for the CEO to hold a meeting after major product launches so employees could ask him questions. This time, however, when someone asked him about Google, the CEO unleashed an awe-inspiring rant against Android and Google.

"We did not enter the search business. They entered the phone business," he said. "Make no mistake. They want to kill the iPhone. We won't let them."

Jobs was so fired up that he continued even after someone else tried to change the topic.

"I want to go back to that other question first and say one more thing," he said. " 'Don't be evil' is a load of crap." Jobs was referring to Google's famous corporate motto. In what came to be known as the "Don't Be Evil" manifesto, Google had said in its filing for its 2004 initial public offering: "We believe strongly that in the long term, we will be better served—as shareholders and in all other ways—by a company that does good things for the world even if we forgo some short-term gains. This is an important aspect of our culture and is broadly shared within the company."

Jobs assured the employees in his Town Hall meeting that Apple planned to deliver aggressive updates that Android wouldn't be able to keep up with. He called the next phone, the iPhone 4, an A-plus update.

What he didn't tell employees that day was that he was also planning to sue HTC over twenty patents.

At the time of the first iPhone's release, Jobs had issued a warning to anyone who thought about copying it. "We've been innovating like crazy the last few years on this, and we've filed for over two hundred patents for all the inventions in iPhone," he said. "We intend to protect them."

Apple wasn't bluffing. Two months after the Nexus One went on sale, Apple filed separate complaints with the U.S. International Trade Commission and the U.S. District Court in Delaware.

"We can sit by and watch competitors steal our patented inventions, or we can do something about it," declared Jobs. "We've decided to do something about it."

The press rightly surmised that HTC was just a proxy for going after Google. As much as it wanted to, Apple couldn't sue Google directly because it would be difficult to make a case against a company that wasn't profiting from the software. When Jobs sat down with Isaacson later that week, he went into a tirade. "Our lawsuit is saying, 'Google, you fucking ripped off the iPhone, wholesale ripped us off,'" he said, calling it "grand theft."

"I will spend my last dying breath if I need to, and I will spend every penny of Apple's $40 billion in the bank to right this wrong. I'm going to destroy Android, because it's a stolen product. I'm willing to go to thermonuclear war on this. They are scared to death, because they know they are guilty. Outside of Search, Google's products—Android, Google Docs—are shit," he said.

Jobs was angrier than Isaacson had ever seen him.

Later that month, Jobs met with Google CEO Eric Schmidt over coffee at Calafia, one of Jobs's favorite cafés, owned by a former Google chef in Palo Alto.

The talk soon turned to Android's similar user interface designs. "We've got you red-handed," Jobs told Schmidt as he once again accused Google of ripping him off. "I'm not interested in settling. I don't want your money. If you offer me $5 billion, I won't want it. I've got plenty of money. I want you to stop using our ideas in Android, that's all I want."

Schmidt disagreed. Android had begun as a start-up in 2003 before the iPhone and before Google acquired it a couple of years later. Its innovations were its own.

Not all of the conversation was hostile. As the two CEOs spoke, a passerby overheard Jobs enthusiastically telling Schmidt, "They're going to see it all eventually so who cares how they get it." The eavesdropper guessed that the two were speaking about Web content. As he surreptitiously snapped photos of Jobs and Schmidt, other people noticed them. "Let's go discuss this somewhere more private," Jobs was heard suggesting before they left.

When photos of the meeting surfaced online, readers and journalists speculated that it was a publicity stunt.

The lawsuit had clearly come from an emotional place. Jobs had felt cheated before in the 1980s when Microsoft grew dominant in the computer industry by licensing out an operating system with a user interface like Apple's. That was partly why the company was careful to protect its turf this time with patents that covered its inventions. Apple was again in danger of ending up in a similar position albeit in another industry with another opponent. This time, Jobs was determined that it would end differently. As the threat of Android loomed, Apple was more prepared for battle.

What mattered now was who controlled the innovations that were rewriting the future of mobile communication. Apple's patents covered an array of designs and technologies particular to the iPhone, from core functionalities to the device's physical appearance to the way the phones were used. One example was the "rubber band" patent, so called because of the way content on the screen bounced to signal when a user reached the bottom. Without it, scrolling stopped abruptly and awkwardly. Though the feature was seemingly minor, it had been one of Jobs's favorite patents. It had inspired him to work on a phone in the first place. Jobs saw the fluidity of the unique action as part of the iPhone's distinguished look and feel, and his anger roiled when its rivals made it commonplace. Apple also had design patents laying claim to the iPhone and iPad's rectangular shape with rounded corners, edge-to-edge glass, and home button.

Apple was determined to bring down competitors who it believed had stolen the game-changing advances developed for the iPhone. HTC was a strategically smart first target. The company was relatively young and had a weak patent portfolio that made it harder for it to countersue. Also, Apple and HTC had no vital business relationship that might be jeopardized in a take-no-prisoners battle. With no reason to hold back, Apple's lawyers were free to attack as ruthlessly as they liked. If they obliterated HTC, then Apple could use the precedent to build its case against the other rivals.

The company's momentum, however, was slowed by its ongoing fight with Nokia. By engaging in two battles at once, Apple had given the court an opportunity to consolidate the cases, which simultaneously slowed the process down and allowed HTC to take advantage of Nokia's defense savvy.

Apple had yet to make much headway in the process when it was embroiled in yet another patent fight six months later. As the turf battle between Apple and the Android device makers heated up, Motorola filed suit against Apple, accusing the company of violating eighteen patents. Apple countersued a few weeks later.

Now Apple was engaged in conflicts on three fronts. And the intensity of the war was showing no signs of abating. Industry research firms were predicting that Android would continue to gain momentum and was poised to overtake Apple in market share in 2011. Leading the charge wasn't HTC or Motorola; it was Samsung, the aggressive South Korean electronics conglomerate with global ambitions.

On the evening after the first iPhone went on sale, a couple of dozen visiting designers from Samsung were dining at a Korean barbecue restaurant in San Francisco called Hanuri when a friend showed up with the device. The phone was locked, so the designers couldn't see the home screen or open the applications. But it didn't matter. They were impressed enough with the sleekness of the device and the elegant ease of swiping their finger to pull up the pass-code screen. They oohed and aahed as they made the gesture over and over again. They had never seen anything like it.

Like the rest of the world, Samsung's executives and designers were awed by the iPhone, and they wanted something similar. Historically, companies chased each other all the time with products that looked the same. If one scored big with a product that was slightly different, others followed quickly with varying degrees of modification. That was how minor companies moved up the food chain. Samsung was no different. The company had access to plenty of top engineers and designers and didn't lack for talent. But their main task was to look at popular products in the market and focus their energies on improving upon them. Part of its modus operandi was to use its manufacturing prowess and its relationships with its customers to follow rivals quickly from behind. When Motorola's Razr phone was all the rage, Samsung's executives demanded that its engineers outdo them with a similar phone that was even thinner than the Razr. The edict was the same with the iPhone.

From Samsung's perspective, there was nothing wrong with being inspired by a rival. Companies were stimulated by each other's products all the time.

The Galaxy S, Samsung's first iPhone look-alike, created an enormous complication for Apple. Unlike HTC, Samsung was a formidable opponent—a family-controlled multinational conglomerate with nearly limitless resources and numerous subsidiaries and businesses from electronics to heavy industries to life insurance. In 2010, the electronics division alone reported revenues of 154.6 trillion won, or $142 billion. The company had a strong portfolio of patents that was only growing bigger.

Although the two corporations were rivals, Apple also happened to be one of Samsung's biggest customers. That same year, Apple had spent about $6 billion with Samsung on microchips, memory chips, and liquid crystal displays. If a legal battle led the Korean company to withhold those components, Apple would be in trouble.

Cook, Apple's reigning expert on the intricacies of the supply chain, was particularly wary of endangering the relationship.

But Samsung was gaining. By the end of 2010, the electronics maker was the fastest-growing smartphone maker. Its 7.6 percent market share was still about half of Apple's, but the company had increased its shipments by 318 percent from 2009. Apple couldn't afford to sit by and let Samsung take the market with a phone that was, at minimum, heavily inspired by the iPhone. The whole premise of Apple's business was that it was a product, design, and thought leader. The unique way it pulled together the overall package was what allowed the company to charge a premium. If it let others copy them, then it would lose its distinctiveness, and its business model would crumble. Apple had to strike before it was too late. In April 2011, it sued Samsung in the U.S. District Court for Northern California, its home court. In a thirty-eight-page complaint, the company accused Samsung of copying the look, product design, packaging, and user interface of its products as well as violating its patents and trademarks.

"Instead of pursuing independent product development," the complaint alleged, "Samsung has chosen to slavishly copy Apple's innovative technology, distinctive user interfaces, and elegant and distinctive product and packaging design, in violation of Apple's valuable intellectual property rights."

To prove the point, Apple's lawyers included side-by-side comparisons of the two companies' phones.

"The copying is so pervasive, that the Samsung Galaxy products appear to be actual Apple products—with the same rectangular shape with rounded corners, silver edging, a flat surface face with substantial top and bottom black borders, gently curving edges on the back, and a display of colorful square icons with rounded corners," it said.

Apple also pointed to similarities in packaging and icons for applications like music, phone, texting, and contacts. The phone icon, for example, was an image of a white handset on a green background at virtually the same angle. The background of its photo app appeared to be sunflower petals, the same flower that represented the photo app on the iPhone.

By this time, Jobs was on what would be his final medical leave, placing Cook in charge of juggling Apple's various interests.

Cook was careful to draw a line between Samsung the supplier and Samsung the rival. When, during a quarterly earnings call less than a week after Apple filed its lawsuit, a financial analyst asked about how the lawsuit could impact its supplier relationship, Cook told him, "We are Samsung's largest customer, and Samsung is a very valued component supplier to us, and I expect the strong relationship will continue. Separately from this, we felt the mobile communication division of

Samsung had crossed the line. And after trying for some time to work the issue, we decided we needed to rely on the courts."

The animosity between the companies would get worse. In the days following, Samsung sued Apple in Korea, Japan, and Germany. It also responded in California with a countersuit. Samsung saw Apple's actions as an attempt to restrain it. In the Korean electronics maker's opinion, the company was merely competing.

Samsung's chairman, Lee Kun-hee, recited a well-known Asian adage to reporters: "A nail that sticks out gets pounded down."

Like Apple, Samsung prized secrecy. The Korean conglomerate had once been described as a fortress from which no information was supposed to reach the outside world. Still, enough facts had slipped through the cracks for Apple to know that it faced a formidable adversary.

Chairman Lee was especially intimidating. Almost seventy, Lee still ruled Samsung with an iron will. In many ways, he was similar to Steve Jobs. Some thought of him as Samsung's "wise emperor." After inheriting Samsung from his father, Lee had led the push to reshape the company into a giant of global communications and electronics. He spoke Korean, Japanese, and English and had taken MBA classes at George Washington University. Sometimes referred to as the King of Korea, he was the country's richest man, with an estimated personal wealth of about $8.6 billion around the time Apple sued Samsung.

According to reports that had found their way past Samsung's walls over the decades, Lee saw himself as a visionary whose calling was to foresee the challenges and opportunities in his company's future. Much of the day-to-day running of the business he left to others. Like Jobs, Lee had earned a reputation for promoting internal competition among his lieutenants. Like Jobs, he was also famous for nurturing a corporate culture of constant and often nerve-racking transformation.

Lee had distilled his beliefs into sayings that he etched into the minds of the thousands who worked beneath him. One motto declared:

"I must be the one to change first in order to survive."

Another:

"Change everything, except your wife and kids."

At an age when other CEOs had retired, Lee would arrive at Samsung's headquarters in Seoul at dawn, or earlier. His lieutenants arrived by 6 a.m.

Lee was the third son of Samsung's founder, Lee Byung-chul, who had started the company in 1938 as an export firm that sold dried fish, vegetables, and fruit to China and Mongolia. The elder Lee had turned to technology to stay ahead of the competition. His drivers transported their wares in a motorized truck, while his rivals relied on oxcarts. As the decades passed, the founder diversified his company's holdings, opening his own factories and acquiring multiple insurance companies. Samsung—the name meant "three stars" in Korean—grew into a sprawling multinational conglomerate, all of it run by the Lee family.

The younger Lee became the chairman in 1987, two weeks after his father's death. By then, he was already encouraging the electronics division to enter the burgeoning field of mobile phones. From the start, copying was key to Samsung's advances, beginning with efforts to reverse engineer Japanese car phones and then moving on to Motorola phones. Determined that his company would eventually add its own innovations to these early imitations, Lee kept up the pressure. By the end of 1994, the company had developed enough generations of mobile phones that the chairman felt confident enough in their quality to give thousands of the latest model away to Samsung employees as a New Year's gift. When many of the gifts proved to be shoddy, Lee responded with a gesture that would be talked about for years to come.

That March, the chairman summoned his board to the Samsung factory where the phones had been made, in an industrial city called Gumi in south-central Korea. The factory's two thousand workers were ordered to don headbands that read *Quality First* and then called to a courtyard where they found the factory's entire inventory—nearly $50 million worth of cell phones, fax machines, and other equipment—piled into a heap near a banner that declared *Quality Is My Pride*. Underneath the banner, Lee sat with his board of directors. At his command, a team of workers smashed the inventory with sledgehammers and then set it on fire. As the flames rose, many of the factory's employees were so chastened that they began weeping.

That was the end of quality control issues at Gumi. After the incident—or as it became known, "the voluntary incineration"—the factory manager tested new models of the phones by hurling them against walls, dropping them from second-story windows, and once, running over one with his car. Usually, he reported, the phone still worked.

In the years since then, Lee had successfully turned Samsung into one of the world's most dominant electronics companies. He had also suffered his share of setbacks, fighting cancer and charges that Samsung had routinely bribed government officials with a slush fund of seven trillion won, or roughly $7.5 billion U.S. Most of the charges were eventually dropped, but after pleading guilty to tax evasion, Lee resigned, publicly apologized for his company's moral and ethical lapses, and went home to his mansion, where he was placed under house arrest until the following year, when the government granted him a full pardon and he returned to Samsung as chairman. Amid the global financial crisis, Lee had been deemed too invaluable to be kept on the sidelines.

The crisis of those days was over. Now Lee's attention was turned toward defeating Apple. He was as strong as ever, still arriving at his desk at dawn, still inspiring terror in those who failed in the quest to help Samsung conquer the world.

When the chairman spoke, a Sony executive once joked, it was like hearing the voice of God.

Between HTC, Motorola, and now Samsung, the battle against Android was heating up. Apple needed to rid itself of any distractions, so it settled its dispute with Nokia. Apple came out as the net loser, agreeing to make an estimated one-time payment of between $600 million and $720 million to Nokia for the use of its technologies in past phones in addition to paying royalties on an ongoing basis for future devices. But that was a drop in the bucket for Apple. The freedom it would gain to pour its energies into fighting Android was far more important. It doubly made sense because Nokia was shifting its phone strategy to focus on Windows Phone, which was protected by Microsoft and left little room for Apple to collect royalties. Apple would now be able to forge ahead into its fight against Android.

The civility between Apple and Samsung dissolved almost immediately. When Samsung sought a court order that would force Apple to disclose the iPhone and iPad models that were under development, Apple accused it of making "an improper attempt to harass Apple." When Apple requested a preliminary injunction in the United States against four Samsung products, the two sides squabbled over the hearing schedule. Samsung, which had more to lose with a faster decision, wanted the hearing pushed back. Apple, which wanted an expedited process, argued for an earlier date. Samsung also sought to disqualify some of Apple's outside lawyers because they had previously done work for Samsung.

Nothing was off the table, even when it risked inviting public ridicule. At one point, Samsung filed an amusing brief in opposition of Apple's motion for a preliminary injunction. In it, Samsung argued that the general design of the iPad was not original to Apple, citing a scene in *2001*, in which two astronauts are watching an interview with HAL on their tablet computers while enjoying a futuristic meal of puréed food.

By the end of April 2012, the two companies had filed more than fifty lawsuits against each other in sixteen courts in ten countries. Some of the most important battles would take place in San Jose, California; Mannheim, Germany; and the U.S. International Trade Commission, which had the power to ban imports into the country.

As the global war expanded, Samsung assumed a tougher stance. Initially, many of the company's executives had called for a quick reconciliation, but over time they were energized by the increased attention they were receiving by the public. In a strange way, Samsung was legitimized by Apple's attacks. The iPhone maker deemed the competition from the Galaxy to be serious enough to challenge.

In an interview with the Associated Press in September, Lee Younghee, Samsung's head of global marketing for mobile communications, accused Apple of "free riding" on its patents. "We'll be pursuing our rights for this in a more aggressive way from now on," Lee said. "We've been quite respectful and also passive in a way. . . . However, we shouldn't be anymore."

Still, when Samsung's president, Lee Jae Yong received an invitation to attend Steve Jobs's memorial service at Stanford the following month, Lee accepted. He was the only Asian executive on the guest list. "Steve Jobs . . . was a demanding customer and competitor, but I grew fond of him," Lee told reporters at Gimpo airport before he departed. "Samsung and Apple should be partners and we should compete fairly and fiercely in the market."

Apple was pushing for preliminary injunctions in most countries with mixed success. It won a ban against the Galaxy Tab tablet device in Germany and certain Samsung phones in the Netherlands. An injunction was also issued in Australia against the Galaxy Tab. All of those would pale in importance next to a favorable outcome in the United States.

On a brisk Thursday in mid-October, attorneys from Apple and Samsung arrived at the federal courthouse in San Jose for a hearing. Presiding over the case was Lucy Koh, a Korean-American judge, whose diminutive frame and feminine voice belied her no-nonsense toughness. In a stark courtroom that would soon become familiar to both sides, Apple's lead attorney asked the judge for a preliminary injunction to stop Samsung from selling its devices in the country. If the court waited until the following year after the trial reached its conclusion, he argued, Samsung would have already done its damage and moved on to selling newer models that would be outside the scope of the lawsuit.

When it was Samsung's turn, counsel Kathleen Sullivan handed out thick binders with 249 pages of slides. "What's happening here is that Apple is going to claim a monopoly right to the design over a rectangular smart phone with a flat surface," she said, going through the slides to show how other phones and handheld devices in the past had included some of the same design features that Apple claimed to be theirs.

"Let me ask you a question," interrupted the judge, holding up a Galaxy Tab and an iPad. "Tell me which one is Samsung and which one is Apple. Tell me which one. I'm not going to show you the back. Can you tell me?"

"Not at this distance, your honor," admitted Sullivan, who was standing about ten feet away.

Koh offered to come down from the bench and bring the two devices closer. "Which one is which? Can any of the Samsung lawyers tell me?"

Sullivan tried to argue that it only needed to raise strong enough questions about the validity of Apple's patents to defeat an injunction. But Koh appeared to disagree as she explored the idea of Samsung's tablets being sold at the expense of the iPads.

"These are just black screens," Koh said. "I'm . . . I'm as close to you as I can get."

Sullivan argued that the judge's test was unfair.

"I don't mean to quibble over the small experiment, but the '889 patent is for the whole device and you didn't show us the back." The counsel kept going. "And the ordinary observer is not limited to the sense of sight, especially not at fifteen feet."

"I don't think this is fifteen feet," the judge shot back, though she obligingly turned the phone over. "Of course you can see the logo. But the back, the shape is very similar. I'm showing all sides now, showing the side, the back. . . . I'm holding them right next to each other. The thinness is very similar; the sloping on the edges is very similar."

The judge appeared to be leaning in favor of Apple, but after deliberating for a month and a half, Judge Koh denied Apple's request for a preliminary injunction. Apple had "established a likelihood of success on the merits at trial," Koh said, but Samsung had raised strong questions regarding the validity of one patent. She said Apple had also yet to prove that it was being irreparably harmed by Samsung's devices.

With the removal of a major threat, Samsung strengthened its resolve to keep fighting.

"We are in for the long-haul," a Samsung executive told the *Korea Times*, adding that its CEO was fully behind the effort. "Samsung doesn't want to be involved with legal issues, but in this case, Apple started it. It's natural for Samsung to defend our bottom line."

Apple's determination to win the battle also remained unshakable in the wake of Jobs's death. The publication of Isaacson's biography had an immense impact as the public read about Jobs's vow to wage war. Though insiders had long been aware of the impending conflict, the best-selling book revealed to the public the intensity of Jobs's animosity toward Google and Android. It was unthinkable not to honor what was essentially the visionary's dying wish to protect his legacy.

The company was also buoyed by a special exhibit that the U.S. Patent and Trademark Office opened, honoring Jobs in the atrium of its museum in Virginia. The presentation included thirty giant iPhone-like displays showing the front pages from more than three hundred filings that bore Jobs's name as inventor or co-inventor.

"This exhibit commemorates the far-reaching impact of Steve Jobs's entrepreneurship and innovation on our daily lives," said David Kappos, undersecretary of commerce for intellectual property. "His patents and trademarks provide a striking example of the importance intellectual property plays in the global marketplace."

In addition to its battle against Samsung, Apple kept attorneys busy as other disputes with HTC and Motorola progressed. In each, Apple nitpicked on the most trivial matters. In a court filing that took aim at one of HTC's complaints, the company complained about a comma.

"Apple denies that its correct name is Apple, Inc. The correct name of Respondent is Apple Inc."

The litigation with Motorola bounced from court to court as lawsuits moved between Delaware, Wisconsin, and Illinois before ultimately being consolidated in Illinois. The impact of this outcome wouldn't be understood until months later.

Over the next few months, Apple and the Android camps fought viciously but with no decisive victory on either side. Apple's injunction against Samsung in Australia was overturned. A German injunction Motorola had won against some of Apple's products was temporarily suspended. HTC also lost one round when the International Trade Commission threw out a complaint.

Apple's disputes centered around two ideas. The first was to protect its innovations. The second was as a defense against more established mobile phone makers like Motorola and Samsung, who accused Apple of violating a category of patents that they were obligated to license to anyone who wanted it. These patents were known as "standard essential" because they had been included in telecommunications standards. In return, companies who owned them were required to offer them under reasonable and nondiscriminatory terms, but Apple was asserting that Motorola and Samsung were unfairly demanding a higher price for them as a way to keep the competition at a disadvantage.

The implications of a win on either side were vast. Like Judge Koh, many observers agreed that Android devices had a similar look and feel to Apple's iPhone and iPad. Yet some of the individual design elements by themselves were not that original. By necessity, a smartphone had to be flat and rectangular in shape to make it easy to hold. Putting rounded corners on it was not particularly inspirational.

It was unrealistic to expect that even a hundred lawsuits could make Android disappear. As Android's market share eclipsed Apple's, it was growing increasingly clear that consumers wanted that choice. The best that Apple could reasonably expect was to slow the competition and force Android device makers to change their products enough to draw an even clearer distinction from Apple's.

Every day that the various lawsuits dragged on was another day that Android was able to blaze ahead. A few months before trial, Apple and Samsung submitted a joint filing to the court, in which both sides sought to eliminate any evidence that could prejudice the jury. Apple's list ran the gamut from an objection to the Samsung logo on the court's video display to an exclusion of Jobs's "thermonuclear" comments.

Samsung was most worried about the influence that Apple's iconic status would have on jurors. It asked the court to exclude product reviews in Apple-specific blogs and fan sites and pro-Apple expert testimony about the company's cultural significance.

In a separate motion, Samsung noted, "Apple's damages expert, Terry L. Musika, writes in his report that 'Apple has built a considerable and at times a cult-like following to all things Apple.' That cult-like following apparently includes several experts who are appearing on Apple's behalf in this case, and may explain why they have cast aside established scientific methods and governing legal principles in favor of slavish adoration of their client and platitudes about its alleged magical and revolutionary products, issues that are of no relevance to the claims and defenses at issue."

Samsung also questioned the objectivity of Henry Urbach, another witness Apple proposed to summon before the jury.

"Mr. Urbach wrote an essay on the design of Apple's retail stores, entitled *Gardens of Earthly Delights*, describing them as "[q]uasi-religious in almost every respect, . . . chapels for the Information Age," Samsung said, pointing out that he also referred to the late Apple founder Steve Jobs as "St. Eve."

In the spring of 2012, the court ordered a final meeting between the heads of the two companies to try to reach a settlement. Cook and Choi Gee-sung, Samsung's vice chairman and electronics chief, met for nine hours on May 21, along with their lead counsels, and for seven hours on May 22 in a federal courtroom in San Francisco. Mediating the session was Joseph Spero, a bow-tie-wearing magistrate judge with a reputation for handling complex cases. Very little is known about what went on in the meetings, but when the talks yielded no settlement, no one was surprised.

A few weeks later, when Cook was asked about the patent disputes at an onstage interview at a technology conference, he told the audience that the legal battles were "a pain in the ass."

"From our point of view it's important that Apple not be the developer for the world," he said. "We can't take all of our energy and all of our care and finish the painting and have someone else put their name on it. We can't have that. And so the worst thing in the world that can happen to you if you're an engineer and you've given your life to something is for someone to rip it off and put their name on it. And so what we want to accomplish is we just want people to invent their own stuff and we don't want to be the developer for the world."

In Korea, independent newspaper *Hankyoreh* wrote about a growing view in the industry that the lawsuits were a way for Apple and Samsung to jointly prevent other competitors from entering the market. Pointing out that neither company had been hurt or helped by any of the cases, it opined that there was no reason to continue the legal proceedings.

The two companies disagreed. Attorneys on both sides prepared for the showdown.

As the July 30 trial date neared, Apple's legal team booked more than fifty rooms at the Fairmont hotel nearby and rented a temporary office across the street. A pantry was filled with refreshments to keep the team alert—coffee, soda, beef jerky, kale chips, Cliff Bars, instant noodles, and everything else in between.

Samsung's team reserved rooms at the Marriott, moving furniture out of some of them to set up a makeshift command center.

War rooms.

The Innovator's Dilemma

Three thousand miles away, in an office in Cambridge, Massachusetts, the professor was watching every twist and turn of Apple's fortunes. Sitting in front of his iMac and scanning the latest news, Clayton Christensen was attempting to divine the future of Apple now that its visionary king was dead. A renowned expert on the rise and fall of corporate empires, Christensen noted how Tim Cook lacked Steve Jobs's exceptional instincts for product development. He observed how rivals were eroding the iPhone and iPad's market share. And though he had long admired Apple's dazzling success, he worried that the company may have finally reached its zenith. Was this the start of an inevitable plunge back to earth?

A professor at Harvard Business School, Christensen had devoted much of his life to developing a theory that explained why so many powerful companies, especially those that reigned through technological prowess, were eventually overrun by competitors making low-end products. Christensen's theory of disruption, laid out in his book *The Innovator's Dilemma*, dissected the forces that made it almost inevitable for these emerging rivals to eventually triumph over corporate giants. In a world where technology was constantly evolving, creating radically different products that opened new revenue streams, the ability to disrupt an existing market—mobile phones, for instance—was crucial to survival. *The Innovator's Dilemma* had influenced everyone from Andy Grove and Michael Bloomberg to Steve Jobs when it was published in 1997. In the years since, as major corporations sought him out for advice, Christensen had increasingly focused his research on how to avoid the trap and stay competitive.

For years, Apple had been considered the exception to his theories as it grew bigger and seemingly more invincible with every product launch. Apple had distinguished itself by putting profits secondary to its primary objective of creating new possibilities in the market and new appetites for consumers. More recently, however, the company seemed to be too focused on beating its rivals. Its products were thinner, more powerful, and more perfect than previous iterations, but these improvements were decreasingly important to many consumers. That created an opening for rivals to enter the market at the bottom end. Apple seemed to be unraveling.

Shortly before Apple began selling its first iPhone in 2007, Christensen famously predicted its failure. "Apple won't succeed with the iPhone," he told *BusinessWeek*. "They've launched an innovation that the existing players in the industry are heavily motivated to beat: It's not [truly] disruptive. History speaks pretty loudly on that, that the probability of success is going to be limited."

Though critics mocked him for being wrong at the time, he was starting to look prescient as Apple's iPhone and iPad began to steadily lose market share to Android-based smartphones and tablets.

But Christensen took no pleasure in being right.

"I truly hope that you won't print me as somebody who has it in for Apple," he said as he paced, his six-foot-eight frame filling his modest-sized office. After suffering from a stroke a couple of years before, he spoke slowly and deliberately. "I'm just trying to learn from them."

Christensen had started his career with the Boston Consulting Group and cofounded an advanced materials firm before pursuing academia. In the 1990s, as Apple teetered on the edge of bankruptcy, he began studying why companies had such difficulty staying successful. The first market he examined was disk drives. What he discovered was surprising. Dominant disk drive makers were losing business to rivals with seemingly inferior drives. Companies that had made fourteen-inch-diameter drives for mainframe computers were supplanted by those that made eight-inch drives for mini computers, which were then replaced by those that made 5.25-inch drives for PCs, and so on. This phenomenon was happening even though the smaller drives had a lower capacity and a higher cost per megabyte.

Upon further exploration, Christensen concluded that the bigger companies failed to recognize the significance of the emerging smaller drives until it was too late. The fourteen-inch-drive companies didn't think much of the eight-inch drive initially because it had a capacity of forty megabytes or less, compared to their drives, which had three hundred to four hundred megabytes. Such small drives were useless to its mainframe customers. But by focusing on their target audience, they overlooked the growing market of other customers who made smaller computers and wanted smaller drives.

He found similar trends in other industries. In retail, most department stores had ignored discount stores until they began seizing market share. While some companies like Dayton Hudson, the owners of Target Corporation, managed to stay ahead by quickly shifting their businesses to compete with the up-and-comers, others like Woolworth's clung too long to their traditional businesses and faded away. In the steel industry, large-scale integrated mills saw their markets increasingly invaded by more efficient, low-cost minimills.

From these examples, he concluded that companies failed not because they were badly managed but precisely because they did everything they were traditionally expected to do. They listened to their customers, studied market trends, and invested in new and improved technologies that they thought their customers would want. They ignored emerging products because they were usually cheap and shoddy and didn't meet their existing customers' needs.

But those very actions paved the way for their own demise because there was a limit to how much they could refine their products before customers lost interest in the incremental upgrades. That was the point at which most customers were satisfied enough with their purchases and were no longer willing to pay for anything more advanced. By the time those companies paid attention to the cheap, new innovations they had initially ignored, it was usually too late. By then, the early entrants had captured new groups of customers and improved their products enough to eat into the established companies' businesses.

"Toyota did not come to America with Lexuses," Christensen explained. "They came with this rusty little compact in the sixties that they called the Corona. And then they went from the Corona to the

Corolla, Tercel, 4Runner, and *then* a Lexus. General Motors and Ford were up here on the integrated-steel trajectory, making big cars for big people."

The challenge for companies was that every new product eventually became established, ready to be disrupted by the next new thing.

Apple had been one of Christensen's case studies. Apple was founded in 1976 as an upstart that upended the industry. While big computer makers were focused on selling mainframe and mini computers, Apple had recognized the potential of personal computers. Its first computer, the Apple I, was little more than a circuit board, "at best a preliminary product with limited functionality," as Christensen described it. But it was successful enough for the tiny company to launch the much-improved Apple II the following year. The 1984 release of the Macintosh cemented Apple's position as a market leader in personal computers.

Then the company hit a wall. Consumers initially paid a premium for the Mac's ability to navigate with a mouse that could point and click on graphical images. Rival companies, however, began selling their own personal computers, loaded with Microsoft's Windows operating system, which provided similar features. The initial models were inferior to the Mac in many ways, but they were good enough, they were cheap, and they got better every day. After years of incremental improvements by Microsoft, the release of Windows 95 brought Apple's software superiority to a nearly complete end as it matched and exceeded the Mac in almost every feature.

In areas where it did try to innovate during this era, Apple made the classic mistake that Christensen also described in his book: It sought input from its customers. The example that Christensen picked up on was the Newton personal digital assistant (PDA). The product had some of the characteristics of a disruptive technology in that it had the potential to take sales away from laptop computers, but rather than start with modest expectations, Apple's CEO at the time, John Sculley, saw it as a key product for the company. He invested many millions of dollars in its development as his team conducted extensive market research, focus groups, and surveys to figure out what consumers wanted. The result was a flop on the grandest scale. A seven-hundred-dollar device with insufficient power and supposed state-of-the-art handwriting recognition that was so inaccurate it was lampooned in a *Doonesbury* comic strip. Had Apple kept its investment and its promises modest, the 140,000 Newtons it sold in the first two years may have been considered a solid start. Instead, the device went down in history as one of Apple's worst failures.

When Jobs returned to Apple, he turned the company back into a disrupter. In Jobs's mind, Apple had become too corporatized during his exile. Its CEOs were professional managers who had little direct involvement in the day-to-day business and were overly focused on profits. Decisions were bogged down in endless meetings and committees. To transform Apple into an innovator once again, Jobs cut non-essential products and put the focus back on making great products. New development projects were protected and overseen by senior managers, often with input from Jobs himself. Ideas were developed, modified, vigorously debated, and sometimes even discarded at the last minute. At the same time, a new process was created so nothing fell through the cracks. Tasks were assigned a "Directly Responsible Individual," known by its shorthand DRI. Apple also stopped depending on market research and surveys. "If I'd have asked my customers what they wanted," Jobs quoted Henry Ford, "they would have told me, 'A faster horse.'"

Instead, decisions were informed by careful observations of user behavior. When a brand-new product was released, they nurtured the category over a period of years rather than expecting a blockbuster right away.

These changes enabled Apple to disrupt industry after industry. The all-in-one iMac was affordable, stylish, and easy to use in a way that computers hadn't been until then. The iPod paired with iTunes transformed the music industry as people began buying songs for ninety-nine cents apiece rather than entire albums. The iPhone defined the smartphone. The iPad, Jobs's last brand-new product, was a device that no one saw a use for before Apple introduced it.

Jobs had a rare ability to maintain a sense of crisis within himself and inside the company. "If you don't cannibalize yourself," he used to say, "someone else will."

But was that still true?

The trends in the mobile device market were starting to look alarmingly like the Innovator's Dilemma scenario that Christensen had painted in his book.

When the first Android phone came out, few people took it seriously. It wasn't attractive or intuitive compared to the iPhone. Its music application wasn't as good as iTunes, and its recently launched Android app market wasn't widely supported.

But Android phones improved. Before long, device makers, from Samsung to Motorola to Amazon.com and Barnes & Noble, launched Android-based smartphones, tablets, and e-readers, giving consumers a broader selection in terms of design and price. Google made the operating system easier to use and expanded the app market. The user experience still wasn't as seamless as on the iPhone and iPad—Apple had the clear advantage of tightly integrated software, hardware, and services—but the technology was becoming good enough to interest consumers. Android was invading Apple's market.

In the three months ending in June 2012, Android phones' market share grew to 68.1 percent from 46.9 percent for the same period a year earlier. In comparison, the iPhone's share fell to 16.9 percent from 18.8 percent. Analytics firms saw a similar trend in the tablet market as well.

Apple was not only losing low-end business, it was also beginning to face competition at the high end from companies like Samsung. Galaxy Note, a 5.3-inch smartphone/tablet hybrid that was initially scoffed at in the United States because of its awkward size, became one of its best sellers in Asia, where people couldn't necessarily afford to buy both a high-end smartphone and tablet. Its initial retail price before subsidies was similar to the iPhone's at around seven hundred dollars.

Would this be a repeat of Apple's experience with Microsoft and Windows?

Christensen's office was located in the depths of Harvard's Morgan Hall, beyond a bright, airy atrium with a fourth-century Roman mosaic floor portraying Tethys, the sea goddess and the mother of great rivers, including the Nile. For a man of his academic stature, Christensen's quarters were surprisingly unassuming and functional. His room was a middle unit, one among many, and his assistant occupied a cubicle in a pod with other faculty support staff. His walls were filled with books, mementos, and a collection of disk drives that he used in his classes to show the progress of technology. A blowup of a *Forbes* magazine cover on which he was featured was propped on a shelf in one corner. In addition to the iMac on his desk, he also owned an iPad, but his phone was a BlackBerry.

Over the years, Christensen had developed ideas about how companies could avoid being disrupted. To explain it, he used a story about helping a fast-food chain increase milk-shake sales. The company had come to him after product improvements based on customer feedback had failed to yield any tangible changes in sales or profits. So Christensen thought of the problem in a different way. He asked customers, "What job were you trying to do that caused you to hire that milk shake?"

What he found was that nearly half of the milk shakes were purchased in the morning as a diversion on long solo commutes to work. These people found milk shakes preferable to doughnuts and bagels because they were less messy and easier to handle. Thus, Christensen's advice for improving the product was twofold: Add chunks of fruit to the shakes to add an element of unpredictability to keep the commute interesting and move the prepay dispensing machine to the front of the counter, so customers could make their purchases more quickly. Understanding the customer's purpose made figuring out the solution much easier.

Christensen explained the same concept in another way. Transporting information during Julius Caesar's life required a horseman and chariot. A railroad handled the same task during Abraham Lincoln's time, while a plane was used when Franklin Delano Roosevelt was president. Now the Internet handled those functions.

"So the job actually hasn't changed at all for two centuries or for two millennia, but the technology that you can employ to get the job done changes quite dramatically. So if you think of the structure of your market in terms of jobs to be done, then as you go through the centuries, you're always looking. Is there a better way to get this job done?"

In Christensen's opinion, two of the CEOs who had an instinctive understanding of this concept had been Sony's cofounder Akio Morita and Steve Jobs.

"Part of the hardest thing about coming up with new products is to figure out a really cool set of technologies that you can implement it with and make it easy, but also figuring out something that people want to do," Jobs had once said. "We've all seen products that have come out that have been interesting but just fall on their face because not enough people want to do them."

Morita, a legendary figure on par with Jobs, was also famous for his preternatural ability to identify consumers' needs. After starting Sony in the late 1940s, he launched hit product after hit product from portable radios and televisions to Walkman music players. In every new business he entered, he disrupted the traditional leaders.

But then Morita left. For a while, the company seemed invincible as it introduced the PlayStation game console and VAIO notebook computer. But it increasingly struggled to come up with revolutionary products. It finally stumbled with the emergence of downloadable music. Sony had started work on a portable digital music player years before Apple came out with the iPod, but it was stymied by competing interests from its music label Sony Music, which wanted to protect its sales. It also chose a proprietary technology over the more widespread MP3 format. Once it fell behind that first crucial move toward digital content, it was never able to regain its former strength.

"Both Morita and Jobs were geniuses at figuring out what they were trying to get done," said Christensen, adding that instincts like that were difficult to transfer.

Sony had gone from being great to being merely good. Could the same happen to Apple?

Christensen was particularly troubled by a few of Apple's tendencies. First and foremost was its policy of keeping its products, software, and services proprietary and closed. Having that kind of control over the user experience was beneficial to companies in the first years after an innovation. But Christensen considered it a handicap as the technology matured and rivals caught up.

Apple had so far escaped the trap by jumping to its next product before the previous product matured. But this worked only as long as the innovations kept coming. If Apple stopped, then its proprietary tendencies would begin working against it.

So far, Apple's reaction to Android seemed to confirm Christensen's concerns. In courtrooms around the world, it was devoting staggering resources in a vicious protectionist fight. At the same time, Apple was making the classic mistake of evolving along the same trajectory rather than redefining expectations.

For the first few years, when Apple launched new models with incremental improvements such as thinness, camera quality, and screen resolution, customers snapped them up even if there was no material change in design. But its devices were maturing to a point where the upgrades were in danger of overshooting what most of its customers desired in a device. The iPhone was a prime example. As the analyst Horace Dediu said in a discussion with Christensen, "We cannot get better-resolution screens than we have today with retina because our eyes cannot perceive any improvements. We don't have improvements that we can do in terms of size because they won't fit in our pockets. We don't have improvements that we can see in terms of memory because, frankly, we cannot consume what's on the device before the battery runs out."

Even if Jobs were still around, this would have been a challenging situation. Many considered Apple to be a disruptor to itself as well as other industries. But it was debatable whether it had truly encountered a disruption like Android in recent times. Apple offered multiple versions of the iPod at various price points, but its intention with the lineup had been to sell them all to every customer for different uses. The iPhone arguably disrupted the iPod business, and the iPad the Mac business, but profit margins for the products were higher than or about the same as the categories they cannibalized.

Tim Cook was a master of spreadsheets, not innovation. Since Cook had taken charge, legions of young MBAs had been hired to help feed the new CEO's love of data crunching. For Christensen, that was a huge red flag.

"When we teach people to be data-driven, we condemn them to take action when the game is over because there's no data about the future," said Christensen, adding facetiously that when he died, he planned to ask God why he only made data available about the past. He put the blame squarely on MBA programs, noting the irony of saying that in a business school office. "In my defense, we have a whole course about disruption and jobs to be done."

At Sony, things began to crumble after Morita's departure in part because it too began relying on professional managers who used data and analyses to help them make decisions about product development. To be fair, it was easier for founders to think radically because they had the moral authority to take such risks. Those who followed in their stead had to justify their actions more than their predecessors ever did. The best way to do that was with evidence. When Jobs left Apple the first time, its CEO, John Sculley, discovered that he was held more accountable for his actions than Jobs.

"You can do things as a founder that you can never do as a hired manager," Sculley recalled.

Managers like Cook also tended to overly focus on profits, the one thing that Jobs downplayed. That, in Christensen's opinion, was what had made Apple exceptional. "Instead of having a profit motive at its core, it has something else entirely," Christensen had noted at the time of Jobs's resignation. "Many big companies like to pretend this is the case—'we put our customers first'—but very few truly live by that mantra. When the pressure is on and the CEO of a big public company has to choose between doing what's best for the customer or making the quarter's numbers . . . most CEOs will choose the numbers."

To Christensen, Apple had two choices. It could open up its operating system and license the technology, which would fuel more innovation and expand Apple's presence in the industry at a greater degree than it could on its own. Or it could come up with another disruptive product category that would forestall it from having to face the Innovator's Dilemma. "The salvation for Apple," he said, "may be that they can find a sequence of exciting new products whose proprietary architecture is demanded by the marketplace."

Publicly, Tim Cook was committed. "Our North Star is to make the best product," Cook said time and time again. "Our objective isn't to make this design for this kind of price point or make this design for this arbitrary schedule or line up other things or have X number of phones. It's to build the best."

Now he needed to prove it.

Boundless Oceans, Vast Skies

As dusk descended on an early summer evening in May 2012, a river of workers streamed out of Foxconn's factories in Longhua at the end of their day shift. Outside the North Gate, a small market lay in wait to entice them into parting with some of the money they had just worked so hard to earn. Food stalls, open from 4 a.m. to midnight, sold noodles, honey cakes, and slices of watermelon and cantaloupe. Stores across the street displayed clothing, cell phones, even lingerie. At Pepsi Smile, the local fast-food restaurant, two dollars bought a drink, french fries, and a fried chicken sandwich.

Soon the streets filled with people. Virtually everyone was a factory worker at Foxconn. The neighborhood, known as East Qinhu Village, was the only refuge where employees could easily relax and enjoy themselves for a few hours. The nearest big city, Shenzhen, was more than ten miles away, but to reach it, one had to transfer several times on the bus. The air was lively as couples strolled by holding hands and groups of friends gathered by the stalls for a bite to eat. Many of them had changed out of their factory uniforms and were wearing T-shirts, shorts, and skirts to stay cool on a warm and humid evening.

By city standards, Longhua was in the middle of nowhere, but for many of the workers who had migrated from the countryside, the area was more urban than anything they had ever known, with its tall apartment buildings, neon store lights, and gigantic factory grounds. The fresh-faced men and women looked vulnerable. The vast majority were under twenty-five. The youngest were sixteen, the legal working age in China. Like Sun Danyong, who had also worked there, many were away from their families and living on their own for the first time. They laughed and chattered like any high school or college students their age.

As the sky grew darker, more young men and women materialized from every direction, dressed in Foxconn's standard polo shirts and dark pants. Foxconn's factories operated around the clock with waves of workers entering and leaving through the gates every few hours. The latest crew was reporting for the evening shift. Among them was a pretty young woman with shoulder-length hair and braces on her teeth.

Though she still could pass for a teenager, Ai Qi was twenty-three years old. Having worked at Longhua for five years, she was already an old hand at Foxconn. As she joined the flood of bodies flowing toward the North Gate, she wondered how much longer she could endure this life. Bracing herself for the tedious hours ahead, she sang the opening to a song. An old pop song, from when she was a little girl.

Today I saw snow drifting through the cold night
With the cold, my heart and mind drift off to faraway places
Trying to catch up in the wind and rain, in the fog you can't tell
the shadows apart

The first lines were sad, but then she reached the chorus.

Boundless oceans, vast skies
Would you and I change?

She sang softly, under her breath. She sang to remember herself, to feel free even as she disappeared inside the machine.

Four months after the *New York Times* accused Apple of exploiting Foxconn's cheap labor, environmental improvements were being made, but life remained as dreary as ever for the factory workers. They woke up, they went to work, they performed the same task repeatedly, and they went home. If anything, many of them felt worse off than before.

The biggest sore point for workers was the restriction on overtime. The issues around pay and excessive work hours were immensely complex. Though Foxconn had pledged to develop a compensation package that prevented workers from losing income from reduced overtime, workers still felt shortchanged because they were making less money than they could have if they had been able to work more hours. Adjusted for inflation, the pay increase that they received to offset the reduced overtime was not as high as it sounded. As Foxconn prospered, the cost of living in Longhua rose. In just two years, the rent for many apartments had more than quintupled.

The workers were at Foxconn to earn as much money as they could; it was their best chance to improve their station in life. Work-life balance might be necessary, but it meant less without financial stability.

Discipline continued to be an issue. On some teams, workers who made mistakes or were late to work were still required to present a self-criticism during meetings. They were no longer openly fined as they used to be, but their pay was sometimes cut. A Hong Kong labor activist group called Students and Scholars Against Corporate Misbehaviour, which regularly interviewed workers, also reported incidents of supervisors humiliating workers and forcing them to write confession letters and copy CEO Terry Gou's quotations. One worker reported that a manager at Foxconn had called him stupid. "The pigs," the manager told him, "can only give birth to the brainless."

The reality was that the workers' plight was born out of powerful, elemental forces beyond the two companies' control.

China was in the throes of massive economic and industrial change. According to the government, 262.6 million citizens, or about 20 percent of its population, had migrated from the countryside to the cities by the end of 2012 to seek better lives. They found themselves caught between the old communist system and their more capitalistic ambitions.

"What is happening now isn't just unique to China . . . but we may go through much more severe pains than other countries," said Ma Ai, a sociology professor at the China University of Political Sciences and Law. He cited three reasons: the culture's historical emphasis on the collective interest over that of the individual, the people's reliance on relationships over law, and the extreme gap between the rich and the poor. Apple and Foxconn played a major role in China's industrial revolution, but fundamental change would take time.

The alienation that the workers felt was a predictable reaction to such a tectonic cultural shift.

No one understood the futility better than Ai Qi.

When Ai reached the factory gates every day, she passed the big trucks waiting to be admitted onto the grounds and punched in her identification card at the employee entrance. After entering her building, Ai punched in her ID again at the check-in area to her workroom. Uniformed guards eyed the workers at each checkpoint. Sometimes they asked to see the contents of bags. Workers were allowed to bring their cell phones, but there had been a time in 2007 when Foxconn had banned phones and portable media players to prevent workers from photographing the products they were making.

Her primary means of communication, Ai carried her phone with her everywhere.

When Ai first joined Foxconn, she was assigned to be a line worker. But within a month, she was sent to a department that made the original molds that were duplicated and used in the factory lines to produce the casings for cell phones, computers, and other products. The job required a greater ability than just assembling products, so the pay was better. The initial work was done by hand using a combination of liquid and solid materials, and usually finished with machines. The workroom was covered in dust. Workers were given dark blue, long-sleeve jumpsuits and masks, but their eyes and hands were unprotected. Of the four women in the department, only two, including Ai, made molds because the other two were afraid that some of the chemicals might cause infertility. The work required concentration, but there was no formal training. Everyone in the department had to learn on the job. When the difficult work became discouraging, Ai gave herself a pep talk.

"You can do this," she told herself. "Just take it one step at a time."

To encourage herself, she looked at a sophisticated mold she'd already completed. If she could make that mold, she reminded herself, she could certainly manage the one she was now holding.

Occasionally, though, the frustration and exhaustion overwhelmed Ai so much that she would smash the mold she was working on. It had taken her a full year before she gained confidence, and even then the stress could leave her shaking and speechless.

Though Foxconn kept production lines for each of its customers separate to protect their trade secrets, her department worked for all of them. On any given day, Ai and her coworkers might work on a mold for a part of a computer, cell phone, monitor, or a printer casing for any one of Foxconn's many customers. Nokia, Motorola, Acer, Dell, Apple . . . the list went on. Most of the time, they didn't even know which customer the molds they were making were for because the product casings were so similar.

The exception was Apple. Ai and the others knew when they were working on Apple's products because they were told so, and the material they used would be of higher quality. More stressed than usual, their team leaders would ask that they be extra careful with their work. The slightest error such as making an angle too round or too sharp or causing a tiny nick in the material was unacceptable. Molds had to be polished until the workers could see their reflection. When Ai was assigned to work on the mold for the Apple logo on the back of the iPhone, it took more than a day to get it right and get the steel-plated mold smooth and shiny enough.

With only one break during her eight- to ten-hour shifts, Ai would be exhausted by the end of her day. As she left her workroom, she'd pass by signs with sayings such as "Prevent crises before they emerge" and "Ducks can make noises, only eagles know how to solve a problem."

For a long time, Ai had felt like she was barely hanging on.

Ai had grown up in a rural village in eastern China, the third of four children. Her family was better off than many others in the area— her father owned a small store—but she had an unhappy upbringing. Her parents fought often, and her grandmother and aunt bullied her mother. Believing that her mother favored her younger sister, Ai left home even before she finished junior high school. She had figured she could save enough money to go back to school.

Her hopes were crushed almost immediately. One of her first jobs was as an operator for an express delivery service in Beijing. She worked seven days a week for six hundred yuan a month, about eighty dollars. When she joined Foxconn, things seemed to look up. Her salary was unchanged, and she occasionally had to borrow money from her friends to survive, but she had more days off. That feeling wore off quickly.

Ai had spent one miserable year living in one of Foxconn's dorms. Unlike a Western school dorm, there was no process for matching up roommates, so she had shared a room with eight to ten other women with whom she had little in common. Some of them would usually be trying to sleep while others were getting up to go to work. She found the room to be dirty, and there was no personal space. Conflicts abounded. Ai shared a bathroom with ten toilets and showers with more than sixty women. She had to stand in line just to take care of her basic needs. There was a community television on each floor, but most of the time the set was broken.

Since then, Foxconn had built newer dorms, where only four to six workers shared a room. But it was competitive to get into one of them, and Ai had grown tired of being among so many people all the time. When Foxconn increased her pay, she rented an apartment. For five hundred yuan a month, she had a living room, bedroom, kitchen, and bathroom all to herself. It was an extravagance to not have a roommate, but the apartment was her sanctuary.

Her home was like a window into her past, present, and future ambitions. Her most prized possession was a laptop computer, made by a Chinese brand Tongfang. A couple of years before, she had scrimped and saved to buy it for 4,700 yuan ($740). She had haggled with the electronics store manager for four hours to convince him to give her a 700-yuan discount. Her full-size bed was the first piece of furniture she had bought for 80 yuan ($12). The pale yellow bookshelf-and-desk unit across from it displayed a photo of her mother. Even though Ai had left home because of her, Ai adored her mother. She had raised Ai, her two sisters, and her brother single-handedly while her father worked. Once Ai was settled at Foxconn, she had invited her mother to come live with her, but her mother was uncomfortable in big cities, and she had to care for her grandsons. Instead, Ai sent her family money whenever she could.

For years Ai had harbored an ambition to become an interior designer. She loved to draw, and this love was reflected in the many pictures that hung on her walls. Next to her desk was one of her proudest works. She had copied an online comic strip of a parable to which she had instantly related. The story compared a man's life to walking on a seesaw from the low end to the high end. Every step that he took threw him increasingly off balance. And as soon as he was close to reaching the higher end, he found himself declining as his weight shifted the balance. The lesson Ai took was that you could try to reach your highest ambitions, but what might make you happiest and most stable was to find the point of balance. The caveat that gave her hope was that it was possible to stay standing at the higher end of the seesaw called life if you were supported at the other end by your friends and family.

In another corner of her room, she had hung a trio of drawings of electric guitars. Each was accompanied by lyrics from the Hong Kong rock band Beyond, who sang many songs about poverty, racism, and other social issues around the world. The band's uplifting message was particularly popular among young Chinese migrant workers. In one of the drawings, she had merged the title of one song with a line from another. "Time flies silently. The unyielding spirit will live with me until the end."

For her, the lines served as a reminder.

"As long as I choose the road for myself, I won't regret it even if I lose everything," she explained. "I would just start all over again."

Her words were heavy with meaning. She had put up these pictures in 2010, a low point in her life. She had never been happy at Foxconn, but she grew profoundly depressed that year. Life at the factory was wearing her down. She felt like her managers worked her "like a dog" and blamed her and her colleagues for anything that went wrong. The scoldings she received when she couldn't finish a task on time became unbearable.

To make matters worse, she had given up a big portion of her life savings to help her brother build a new house in preparation for his marriage. She had sent the money willingly, but her diminished savings was still a blow. She had harbored ambitions of going to university one day. The money was supposed to pay for that education and all the hopes she pinned to a degree. Foxconn offered educational as-sistance for technical programs and junior colleges, but she wasn't in-terested in programs that could help her do her job better. She wanted to study sociology.

For a while she felt like she was going nowhere. She had textbooks on her bookshelf to help her prepare for the university entrance exam one day, but her supervisor had warned her that she would lose her job if she was distracted by her studies.

She knew she would never make enough money at Foxconn to feel financially stable, but without a high school or college diploma it would be difficult to find another job that paid as well. She would lose her seniority and have to start from the bottom again.

She also felt unattractive and lonely. Her hands were ruined by her work, and her coworkers teased her about her teeth, which at the time were still crooked. They told her that her smile made her ugly. Even worse, she had discovered that a young man she'd been dating in the city of Shenzhen was two-timing her with a serious girlfriend.

"I knew my love would never be returned," she explained. "So I stepped out of the relationship. I cannot be immoral."

Ai received no support from her family and no understanding. Her mother wanted her to move back home and marry someone in the village. She kept trying to fix her daughter up with men Ai had never met. When one of these would-be suitors called her, Ai listened to him distractedly, letting him talk as she surfed the Web on her laptop.

The emptiness was crushing. Every day was the same. She woke up, put on her uniform, walked through the North Gate, and lost herself working through another day of routine, repetition, and end-less pressure from her bosses. At the end of her shift, she filed back through the gate and out into the city, returning to her apartment to sleep alone.

What was the point of it all?

Foxconn did what it could to make its workers' daily lives seem more like a community. The Longhua campus encompassed dormitories, canteens, a fire brigade, hospital, and an Olympic-sized swimming pool. A company town included restaurants, ATMs, a grocery store, and an Internet café.

A free weekly newsletter called "Foxconn People" updated workers on the community's goings-on and reflected the family-like image that Foxconn was trying to convey. A June 2012 issue included a front-page story about a sports meet that Foxconn had held. In an adaptation of a similar event that is held in Chinese schools, about three thousand workers from three Foxconn facilities competed against each other in track and field contests such as running and jumping for a chance at winning some of the 80,000 yuan ($12,600) in prizes that the company handed out that day. "The three-day games not only offered a chance for the athletes but was also a visual spectacle for Foxconn's sports fans," the write-up said.

Other articles included a piece that lauded Foxconn's first disabled hire, a twenty-two-year-old known as "Tiny Girl" who was hired at a job fair. The girl, who split her time between photocopying and storekeeping at the company's Chengdu plant in Southwest China, stood at just four feet, three inches tall since a childhood accident caused a problem in her spine. "I sent my resume to other companies, but I stopped by Foxconn's booth because I thought the interviewers looked very kind," the article quoted her saying. The newsletter took pains to note that she was being paid the same as other employees and that Foxconn had given her two different jobs "to reduce her boredom and let her make more friends."

At the bottom of the same page was a profile of a Chengdu plant manager called Brother Ping, which praised his hard work and his love for his coworkers. The article noted how he used his own money to buy the workers helmets before the plant was fully up and running and how he personally escorted sick workers to the hospital. Brother Ping made time to organize leisure activities for the workers such as karaoke and hiking, the profile said, even though he worked so hard that he sometimes forgot to eat his lunch of instant noodles.

Another section of the newsletter included a story about a member of Foxconn's "Concern and Love for Workers Team," who performed magic tricks to relieve workers' stress, and columns that offered advice. For a Father's Day gift, the newsletter suggested a cell phone, watch, laptop computer, or camera. To avoid shoe odor in the summer, it recommended crushing a camphor ball and sprinkling the powder evenly inside the shoes.

One of the most popular features was a half-page matchmaking section where single workers submitted blurbs and photos in the hopes of finding a partner. "Sometimes I'm lively and sometimes I'm gentle and quiet. I believe in fate and I hope to make friends from all over the country," wrote one twenty-year-old girl nicknamed Hulala, whose ideal match was someone who was "mature, steady, filial, and ambitious." A thirty-year-old man, who called himself "Cat eats fish," described his personality as mature, steady, ambitious, and caring. "I look forward to starting a sweet home," he said. He was looking for a potential wife who was kindhearted and filial.

But no matter how hard Foxconn's newsletter tried to reinforce the happy image of its facilities, it didn't take long for workers to understand that it was just a façade. Ai had looked forward to reading the paper every week when she first started at Foxconn, but she didn't read it anymore.

"They always write about how perfect Foxconn is, how obedient the workers are, and how good the treatment is," she said. "They never talk about the real labor conditions." A small article on one of the inside pages proved her point. It praised Foxconn's Workers Association for doing a good job. Anyone who had been at Foxconn for a while knew the real truth: Its personnel were appointed by the company and were puppets.

After spending years at Foxconn, Ai had grown despondent about her prospects. In her darkest moments in mid-2010, she had even considered suicide. "I felt like I had no hope," Ai said, looking back. She had asked herself, "Must I live like this for my whole life?'"

Around that time, both Apple and Foxconn were enjoying some of their strongest growth. Hon Hai, the official company name of Foxconn, consistently ranked at the top of *Fortune*'s annual ranking of Taiwan's largest companies. In 2011, it reported revenues of $117.5 billion. That was more than many of its own customers, including

Apple, Sony, and Microsoft. Around the same period, Apple reported a yearly profit of $26 billion as revenues grew 66 percent to $108 billion.

The China market was booming as a solidifying middle class hungered for Apple's products. The company's revenues there quadrupled to $13 billion.

When Morgan Stanley analyst Katy Huberty asked about Apple's business in China in the quarterly earnings conference call, Cook was eager to provide details.

"The China progress has been amazing," he enthused. "Certainly in my lifetime, I've never seen a country with as many people rising into the middle class that aspire to buy products that Apple makes. I think it's an area of enormous opportunity, and it has quickly become number two on our list of top revenue countries . . . the sky is the limit there."

The Chinese perspective was far less rosy. In 2010, China had passed Japan to become the world's second-largest economy after the United States as its gross domestic product increased 9.8 percent to $5.88 trillion. But there was a huge gap between the wealthy and the poor. The average per capita income in 2011 was 23,979 yuan ($3,780) for urban Chinese and 6,977 ($1,095) for rural Chinese, according to the Chinese government. About 150 million Chinese lived on less than a dollar a day. With the iPhone starting at about 5,000 yuan and the iPad starting at about 3,000 yuan, they were big purchases even for the urban middle class. Some young people received them from their parents as a reward for performing well in school, though many had to cut their daily expenses and save up for them. In extreme cases that were widely reported, two high school students were so desperate to buy an iPhone and iPad that they sold their kidneys.

For factory workers, these luxuries were completely out of reach. Ai and the others could only watch as the products they made passed through their fingers into shipping containers that were sent around the world to consumers whose lives they could only dream of.

What saved Ai from giving up on her life was not any improvement Foxconn made, but her music. She had found special solace in the songs of Beyond. From a small speaker on her desk in her apartment,

Ai often listened to the soft voice of Wong Ka-kui singing "Boundless Oceans, Vast Skies," the song she sometimes sang on her way to work.

Not long after writing and recording Ai's favorite song, Wong had fallen from a stage and died. In the decades since, his death had invested his songs with an unshakable power. For Ai, Wong's early end was a reminder that none of us knows how much time we've been allotted. On her days off from the factory, she blasted "Boundless Oceans, Vast Skies" while she cleaned her apartment. At night, she listened through headphones so she could envelop herself in the lyrics.

Still I am free, still I am independent
Always loudly singing my song, traveling thousands of miles

Forgive me this life of uninhibited love and indulgence of free-
dom
Although I'm still afraid that one day I might fall
Abandon your hopes and ideals, anyone can do
I'm not afraid if someday there's only you and me

Ai loved the spirit of the song. On her wall, she had hung another poster emblazoned with another line written by Wong.
It doesn't matter even if the future road is long and unknown.
Ai had regained her belief that she would somehow find a way forward. She began saving money again. She was making about 4,000 yuan per month, or about $630, excluding overtime. Her annual salary of about $7,800 was the equivalent of what an average hardware engineer in the United States made in a single month. But had she been an ordinary line worker rather than a mold maker, she would have been making 3,000 yuan at the most, including overtime. With her new savings, she could now afford braces to straighten her teeth, and she learned not to worry about people making fun of them, too. It was during this time that she had also bought her precious computer and acquired an acoustic guitar.

Now, when she got home from work, the first thing she usually did was boot up her computer. After taking a shower to cleanse away the grime from her work, she would check her messages and read the news even before she ate. It was her one link to the outside world. Ai spent her days off practicing her guitar. The song she was currently learning was called "Has Anyone Told You?" by the musician Chen Chusheng, about young people who were struggling to pursue their dreams in the city. Ai hadn't completely given up on her hopes to go back to school, but her immediate goal now was to leave Foxconn and start an organic farming business that would provide jobs for other young workers who felt the same despair she had. She chose organic farming because she thought there might be healthy demand from China's wealthy elite, who were worried about feeding their children tainted produce.

She also began educating herself on workers' rights. When colleagues got injured at work, she encouraged them to seek compensation, though many of them didn't because they were afraid of retribution from their managers. One of her heroes was Pun Ngai, a sociology professor at the Hong Kong University of Science & Technology and the founder of Students and Scholars Against Corporate Misbehaviour. After spending eight months working on a factory line in 1995 and 1996, Pun had written about how global capitalism, state socialism, and familial patriarchy conspired to exploit young rural women who were put to work in factories for a few years before being recalled home to marry. While writing passionately about the oppression that these young women faced, she also foretold the coming of a silent social revolution, in which the workers would rise above their lowly status.

Ai's experience mirrored that of the girls that Pun had described in her book. But Ai was determined to map her own destiny. To preserve her independence, she avoided going home for Chinese New Year.

"I think if I give up my dream, get married, and have children like traditional women, my life will become meaningless," Ai said on a warm evening in May on her day off, as she relaxed in a pretty black and white polka-dot blouse, sipping an orange drink at Pepsi Smile. Though she complained about the sad state of her hands, one couldn't tell from looking at them. Like many young Chinese women, she wore no makeup, but her fingers had been carefully manicured. A small vanity ring adorned her pinkie.

Later, as she walked through town, she passed by a large group of young men gathered around a television playing in the window of a small electronics shop. They stared at the screen, their eyes empty with boredom.

"They have no goal in life," Ai said, glancing their way, "and they spend their money on silly things."

She understood their aimlessness because she had fallen into that trap as well. Pitying them, she walked on.

Fight Club

The trial between Apple and Samsung was scheduled to open in San Jose on July 30, 2012. In the weeks leading up to then, the courts kept encouraging the two sides to settle.

"Can't we all just get along here?" asked Koh at a June hearing, suggesting they work with a mediator. "I will send you with a box of chocolates, whatever."

Surely they both stood to lose too much if they gambled and put their fate in the hands of a jury. But as the opening day drew closer, there were inevitable signs that war was coming. Just as countries amass their troops in preparation for conflict, Apple and Samsung mobilized battalions of lawyers. As the trial neared, almost eighty lawyers filed notices of appearance. Some represented other technology companies, including Motorola, Qualcomm, and Intel. Between the two companies the lawsuit encompassed sixteen patents, six trademarks, five claims about the design and appearance of products, and antitrust allegations spread across thirty-seven products.

The heart of the case, however, was a very simple question—did Samsung copy the iPhone and iPad?

Though the dispute was just one of a multitude of lawsuits between Apple and Android device makers around the world, this was the first major case to go to trial. Together, Apple and Samsung accounted for 55 percent of global smartphone shipments and more than 90 percent of the market's profits, according to ABI Research. The jury's verdict could have profound ramifications for the competitive landscape in the industry, especially since courts in other countries and jurisdictions would be following the proceedings closely to help inform their decisions. In a high-stakes battle like this one where the final outcome was likely to be some kind of all-encompassing settlement, a definitive win for either side would give it immense advantage in any ongoing negotiations.

In addition to journalists, people following the case included patent experts, regulators, antitrust litigators, policy makers, and standards-setting organizations as well as other companies with patent interests. Even legal historians were following what they considered an epic battle that could be as significant as Thomas Edison's patent claims over the transformative invention of the incandescent lightbulb. Similar to Apple, Edison had not invented the first lightbulb. He was the developer of one of the first commercially successful bulbs, but the race was too close to declare a clear winner. Still, after a decade of legal challenges among the many patent holders, Edison had emerged victorious in the United States, giving him control over the entire market.

"Innovation has a broad-based impact on the economy," said Lea Shaver, a law professor at Indiana University. "Smartphones have become the technological terrain that companies are fighting to grab a piece of through their patents." She pointed out that Edison's win slowed down the competition, making the lightbulb less affordable to the average American for a long time.

In Germany, a patent litigation consultant named Florian Mueller kept one of the most comprehensive accounts of patent cases involving Apple and the Android camp in a globally read blog called FOSS Patents. FOSS stood for Free and Open Source Software. Mueller had spent much of his career advising in licensing deals and partnerships between German and American companies before focusing on intellectual property and competition issues. He filed lengthy articles that parsed the latest developments in trials around the world and laid out the broader implications.

Mueller saw the Apple-Samsung battle as an important test of the patent system. Samsung could argue details about how other devices before the iPhone were also rectangular with rounded corners, but in Mueller's mind, the iPhone had unquestionably redefined the way the user interacted with devices.

"If the patent system fails to protect Apple, it would not just be a failure for Apple as a litigant, but I think it would also be a big-time failure for the patent system as a whole," he said from his home in Munich. "If it cannot protect an undisputed innovator, game changer, revolutionizer, and disrupter like Apple, who is it ever going to protect?"

The public war raged on television, where Samsung ads painted Apple's devices as inferior to its own. In one, presumptive Apple fans camped out in front of a store in Austin, Texas, waiting for the latest iPhone—presumably the iPhone 4S—to go on sale. As they watch a video of someone in London unboxing his new model, one of them groans, "Aw, that looks like last year's phone!" When one of their hipster friends shows up with a Samsung Galaxy SII, boasting about the free turn-by-turn navigation feature in his phone, the Apple customers can't help but express envy.

"We just got Samsunged!" says one. The ad ended with the slogan "The next big thing is already here."

The snarky ads, first aired during the Super Bowl, were notable for their outright animosity toward Apple. If there had been any doubt whether Samsung was ready to open fire on one of its biggest component customers, none existed now. Samsung's attempt to seize some of Apple's cachet was obvious, but viewers loved the bold attacks on Apple's arrogance. The audacious commercials chipped away at the iPhone's cool image.

Still, ahead of the trial, Apple appeared to be winning on the public relations front, so much so that when Samsung unveiled its Galaxy SIII model, some critics charged that it was designed by its lawyers, just different enough to avoid Apple's ire.

"Our change in smartphone design is part of a five-year plan," protested a Samsung executive at the time, "not a sudden turnaround."

Samsung's lead design executive, Lee Min-hyuk, known as "Midas" for his success with the Galaxy series, took personal affront when Reuters asked him to respond to Apple's accusation that Samsung's devices were a blatant steal.

"I've made thousands of sketches and hundreds of prototype products. Does that mean I was putting on a mock show for so long, pretending to be designing?" he asked. "As a designer, there's an issue of dignity . . . I'm the one who made it."

Tensions between the two competitors crackled as both sides sought the upper hand in a series of last-minute legal maneuvers. Apple was seeking $2.5 billion in damages and an injunction to prevent Samsung from selling any devices that infringed on its patents. The company asserted that Samsung should have considered more ways to differentiate their products such as by exploring a non-rectangular shape or choosing a more "cluttered appearance."

Samsung was not only questioning the validity of patents that Apple accused it of violating; it was also asserting that Apple was stealing Samsung's mobile communications technology to make the iPhone and iPad work. Samsung was willing to license its technology, but it wanted 2.4 percent of the full price of the iPhone rather than the half penny per unit that Apple was offering.

Unlike in many other countries where judges well versed in patent law handed down the decision, a jury of ten ordinary citizens would decide the U.S. trial. In a complex case swirling in arcane and often dreary technicalities, the ability to tell a coherent, persuasive story was crucial. To lay out their competing narratives as clearly as possible, both sides needed to get the building blocks of their cases admitted into evidence.

Both parties filed mountains of motions and countermotions. Samsung requested that both companies be referred to by the neutral term, "claimants," even though Samsung was technically the defendant and Apple the plaintiff in the primary claim. It also wanted to switch its seats to the plaintiff's table while it was arguing its counterclaims against Apple.

"Equal treatment of the parties with respect to where they sit while presenting their affirmative case . . . ," the motion went on and on, "will mitigate any prejudice to Samsung that may result from Apple being in closer proximity to the jury throughout the trial."

Table location was a big deal to lawyers because it sometimes made a difference. In courts with no rules about seating, trial teams would arrive as early as possible to claim the side closest to the jury.

U.S. District Judge Lucy Koh approved Samsung's request to be called a claimant but rejected the notion that the lawyers should play musical chairs.

This flurry of requests paled next to the blizzard of motions that roared forth as the two sides fought over the wording of the jury instructions. In a 361-page joint filing, Apple called Samsung's proposal "long and convoluted," while Samsung demanded the insertion of the word *alleged* in claims that Apple had yet to prove. Samsung also separately submitted seven hundred questions spanning forty pages that it wanted the jury to answer in its verdict. Apple, meanwhile, had requested only forty-nine questions.

Each side's strategy was becoming evident. Apple wanted to simplify the case and focus on the big picture. Samsung wanted to immerse the jury in the nitty-gritty.

In the days leading up to the trial, both sides worked furiously with their first witnesses. Lawyers playing the role of opposing counsel hit them with every question they could think of. What is this? they'd ask, showing a slide or a document. Why are you saying something differently than what you said ten weeks ago in the deposition? Explain the inconsistency. Both sides prepared with an eye toward an appeal.

Judge Koh radiated scrupulous impartiality. The judge was a former intellectual property attorney and a 2010 Obama appointee who had an impressive capacity for details. Equally tough on both sides, Koh backed up each of her rulings with unshakable citations from past cases and procedural rules, seasoned with a bracing dose of her own pragmatism. She denied Samsung's request that it be allowed to tell the jury about Steve Jobs's vow to go "thermonuclear." She forbade Apple from insinuating in front of the jury that Samsung was avoiding taxes in the United States.

From the beginning, Koh empathized with the jury, whose members would be paid just forty dollars a day to show up in court for weeks and sift through an ocean of evidence. Koh later gave them a ten-dollar-per-day raise, but it was still a pittance compared to the hundreds of dollars in hourly rates being paid to the lawyers and expert witnesses.

"I think that's cruel and unusual punishment to a jury, so I'm not willing to do it," she told the attorneys as she ordered them to narrow their claims to a more manageable load. She also imposed limits on how long the trial could drag on. Each side could introduce no more than 125 exhibits. They each had a total of twenty-five hours to present their case.

For a trial entangled with global complexities, it was not much time. Samsung's lawyers, eager to escort the jury deeper into that complexity, were now backed into a corner. Once the judge ruled, they had two choices. They could focus their case more tightly, relying on fewer witnesses and tailoring their evidence to fit the available window. Or they could rush, presenting everything in fast-forward, all the while knowing that such acceleration might overwhelm the jury.

The future of mobile communications was at stake, and the clock was ticking.

On July 30, 2012, a warm summer day, the trial opened at the Robert F. Peckham Federal Building courthouse. Hordes of reporters and lawyers waited in lines that snaked around the drab, bureaucratic five-story building. Everyone, including the seventy-four citizens who had been summoned as potential jurors, was required to enter through a metal detector, where they removed their watches and belts and briefly surrendered their electronic gear, including a slew of iPhones and Galaxies. The room where the trial was going to take place was bigger than Judge Koh's usual courtroom, but it was still far too small to accommodate the masses lined up outside. The first reporters, who had arrived as early as 7 a.m., snagged the handful of press seats, but everyone else was ushered into an overflow room where monitors were set up.

The morning started with the airing of outstanding issues before the judge. In the tsunami of pretrial motions, Samsung had already protested the use of "gratuitous" images of Steve Jobs that appeared in five slides Apple wanted to show during opening arguments. Judge Koh had overruled the objection, but now Samsung's lawyers insisted on raising the issue again, asking the judge for clarification on a photo from the Steve Jobs exhibit at the U.S. Patent and Trademark Office.

The image of the fallen CEO, honored like a saint at the high temple of patents, was too much for Samsung to endure.

"That's completely prejudicial, Your Honor," said Charles Verhoeven, the hard-charging lawyer leading Samsung's team. "They've got a picture of him there. You know he's going to influence the jury and prejudice the jury into this popularity contest issue."

Verhoeven, known to colleagues as Charlie, was one of Quinn Emanuel Urquhart & Sullivan's best patent attorneys, having successfully defended Google and Cisco. In 2010, he and his team were named "IP Litigation Department of the Year" by *American Lawyer* magazine, beating out both of Apple's law firms. In law school, Verhoeven had considered dropping out because he didn't want to be a hired hand. But he had changed his mind when he was introduced to litigation.

"Where else, outside of professional sports," he said in an interview, "can you have a battle with an adversary and judge or jury?"

Verhoeven was exhausted after a grueling year in which he had tried seven cases. He was also suffering from sciatica and in a great deal of pain. He had to sit on a stool and take regular cortisone shots, but he refused to allow these factors to slow him down. Most of the people in the room had no idea that there was anything wrong with him.

On the other side of the aisle, with his large frame and neatly combed silver hair, Apple's lead counsel, Harold McElhinny, projected a folksy, disarming air, neither too brainy nor too slick. In his spare time, he liked to travel along the Silk Road in far-flung places like Uzbekistan. His terrifyingly dry sense of humor and demanding style were legendary at his firm Morrison Foerster. According to one story that circulated about his early career, when a junior attorney, trying to make conversation with him in an elevator on a Friday afternoon, asked about his plans for the weekend, McElhinny had supposedly replied, "That's none of your fucking business!"

It was said in such a way that the poor attorney wasn't quite sure if he was kidding or not.

McElhinny had a sharp technical mind and vast experience in litigation, but in court he came across as exceedingly personable. Like the television detective Columbo, McElhinny projected a perplexed air as he made the case that it was just a photo of the patent office and that Apple had no influence over what it displayed.

"We didn't put the picture of Steve Jobs on there."

The judge promptly overruled Samsung's objection.

Once the judge had dealt with the influence of Jobs's ghost, the court was ready to pick a jury. The pool of prospective jurors was a quintessential cross section of Silicon Valley. Several of them had ties to Apple or Google and many had read Isaacson's biography of Jobs. When the lawyers asked the pool what phones and tablets they owned, one potential juror—a Google employee—named multiple Apple and Samsung phones and tablets as well as the Amazon Kindle and Barnes & Noble Nook readers.

"You're good for the economy," Judge Koh told him, eliciting laughter from the observers.

In response to the attorneys' questions, five potential jurors revealed they had patents issued to them. One man, an engineer who was ultimately dismissed, was named in 125 patents related to physics, robotics, and semiconductor manufacturing. When the judge asked if the pool had read anything about the lawsuit, most raised their hands. What was also striking was the international nature of the jury pool. At least eight of them were born outside of the United States. Several had advanced degrees and technical backgrounds. One was a retired San Jose resident who had worked in the hard-drive industry for more than thirty-five years.

By mid-afternoon, the two sides had picked their jury—an eclectic group of seven men and three women, including a social worker, an engineer, a bicycle store manager, and a young video game enthusiast. The hard-drive veteran was chosen to be the foreman. One of the women dropped out right away after discovering that her employer wouldn't pay her while she was away, so the trial proceeded with nine jurors. Two were originally from the Philippines and one was Indian American. Only three owned smartphones. Most owned just a basic phone by LG or Samsung. Only one had an iPhone. The video game fan didn't own a phone at all.

At the end of the day, after the jurors had been sworn in and sent home, Samsung attorney John Quinn asked the judge for one more change.

"Your Honor, there is a sign outside the elevator and downstairs that says 'Apple versus Samsung,'" said Quinn. "It also should say 'Samsung versus Apple.'"

To a layperson, the request would have sounded ridiculous. But to the Samsung team, it was a crucial distinction in how the case would be framed.

The judge acquiesced. The sign was quickly amended. For the rest of the trial, the case would be billed as "Apple v. Samsung, Samsung v. Apple."

Opening arguments began the following morning. Judge Koh issued stern instructions. "You can only say what the exhibits will show, what witnesses will testify to. There should be no argument, no inferences, no arguing the law, and if you do that, I'm going to stop you in the middle of your opening and ask you to please stop arguing the case," she warned. "Please don't cross the line."

McElhinny spoke first about the risks that Apple had taken in developing the iPhone and how Samsung had shamelessly copied it.

"The evidence will show that Samsung had two choices: It could accept the challenge of the iPhone, it could create its own products, it could innovate, it could come up with its own designs, it could beat Apple fairly in the marketplace. Or it could copy Apple," he said. "As we all know, it's easier to copy than to innovate." McElhinny promised to show confidential documents to prove his case.

When he was finished, WilmerHale's Bill Lee took over to present Apple's defense against Samsung's countersuit. Like McElhinny, Lee had an easygoing, obliging demeanor that disguised a keen intelligence and tenacity. As one of the nation's most admired intellectual property attorneys, he had a long list of wins in his nearly four-decade career. His cross-examination of witnesses in cases like the Broadcom-Qualcomm dispute was legendary. A former college athlete, he still ran 125 miles a month. Even during this trial, he stayed in Palo Alto so he could go jogging in the early mornings.

Lee stressed that Samsung's patents concerned older technologies that played only a very small part in Apple's devices. Some of these patents were standard essential, the category of patents that had to be made available to everyone at fair, reasonable, and nondiscriminatory terms. Samsung, Lee argued, was violating that requirement.

Verhoeven's defense of Samsung was straightforward—Apple's patents weren't valid, he said, because there were plenty of examples of similar designs and features before the iPhone and iPad appeared on the scene.

"What the evidence will show is there's an evolution in technology, in smartphone technology, and the evidence will show that as the guts of these phones got more sophisticated and more sophisticated, you could do more things. It's not just Samsung," he said. "The entire industry moved this way."

Verhoeven agreed that the iPhone was an inspiring product. "But being inspired by a good product and seeking to make even better products . . . is called competition. It's not copying. It's not infringement. Everybody does it in the commercial marketplace," he said, adding that Samsung had invested $35 billion in research and development to help build its smartphones. "Samsung is not some copyist, some Johnny-come-lately who's doing knockoffs."

In a move designed to capture the jurors' interest, Apple opened its case by calling to the stand three of Apple's high-ranking officials. Tim Cook was not among the witnesses, but few dwelled on his absence. Samsung's Chairman Lee wasn't planning on testifying, either. Neither executive had played a direct role in the case. Either way, the throngs of spectators were mesmerized by the realization that Apple was about to break its sacrosanct vows of secrecy and reveal its inner workings in open court.

The first witness was Christopher Stringer, a veteran industrial designer who had worked at the company for nearly as long as Jonathan Ive. He was testifying instead of his boss. Ive was in London, attending a Royal Academy of Arts creative industries event with the Duchess of Cambridge, the prime minister, and other notables.

Stringer looked like no one else at court. Dressed in a light suit with long rock-star-like hair and a neatly trimmed salt-and-pepper beard, he fulfilled every expectation of how a hip designer should look. Some people compared him to Viggo Mortensen's character Aragorn in *The*

Lord of the Rings movies. If Apple's intent was to dazzle, it had suc-ceeded. Even before Stringer spoke a word, his physical presence was already testifying to the company's legendary, subversive cool. In ef-fect, the designer was the first exhibit, another Apple product, poured from a mold that had been polished to a sleek and seemingly effortless elegance.

The journalists in attendance, normally blasé about courtroom the-atrics, could not help swooning. Was his suit ivory or beige? Cotton or linen? And how about that beard?

As Stringer glided into the courtroom and settled into the witness chair, his studied perfection instantly affirmed the heart of Apple's case: Of course Samsung was desperate to copy us.

What he said was equally engaging as he talked about how Apple's products were conceived around a kitchen table that Ive had set up in the industrial design studio. "It's where we're comfortable. It's where we are most familial," he said, providing a glimpse into the heart of Apple. "It's a brutally honest circle of debate." Stringer, who was born in Australia and partly educated in Britain, spoke with a subtle accent that could have been Australian or British or something in between. That too added to his suaveness.

Stringer's testimony focused on how the team had designed the iPhone. The point was to emphasize the bold risk Apple had taken in creating such a unique device that was unlike any other in the mar-ket. But even as he acknowledged Steve Jobs's early doubts about the iPhone, Stringer couldn't help but wax poetic about how their labors had resulted in such a wondrous machine.

"It's very simple. It was the most beautiful of our designs," he said, sounding exactly like Ive. "We sometimes don't recognize it instantly. It may take some energy and adding detail. But when we realized what we had, we knew it."

When asked about Samsung's Galaxy devices, Stringer didn't mince words.

"We've been ripped off, it's plain to see. . . . It's offensive."

Apple's next witness was marketing chief Phil Schiller, who offered further evidence of how the iPhone's and iPad's status was revolu-tionary. Though he was a regular presenter at Apple's product launch events, Schiller looked less comfortable on the witness stand in an ill-fitting dark suit. None of Apple's executives wore such formal wear to work.

If he lacked the charisma of Stringer, Schiller made up for it with enthusiasm. The jury paid close attention to his testimony as he revealed more secrets about how Apple was looking for another category it could reinvent after the success of the iPod.

"People started asking, well, if you can have a big hit with the iPod, what else can you do? Make a camera, make a car. Crazy stuff," Schiller recalled. "With the iPod, we realized that if anything were ever to challenge the idea that you have all your entertainment in your pocket—your movies, your photos, your music—it might be the cell phone."

Schiller talked about how Apple had invested more than $1.1 billion in advertising to help turn the iPhone and iPad into a success and how shocked he was when he saw Samsung take advantage of its innovation with nearly identical imitations.

"When you copy or steal the idea of one company's product," he said, "now you're trading off of all that investment in marketing, all that good will we've created with customers."

Apple concluded its executive testimony with mobile software chief Scott Forstall, who looked relaxed as he spoke about how his engineers secretly developed the iPhone operating system. In homage to the project's code name, Project Purple, their building was dubbed the Purple Dorm.

"We started with one floor. We locked the entire floor down," he said. "We put doors with badge readers. There were cameras. I think to get to some of our labs, you had to badge in four times to get there."

Apple fans had long heard of how new products were developed under lockdown, but to hear Forstall confirm the legend in such detail was thrilling.

"People were there all the time. They were there at night. They were there on weekends. You know, it smelled something like pizza," he said. "In fact, on the front door of the Purple Dorm, we put up a sign that said 'Fight Club' because the first rule of Fight Club in the movie is you don't talk about Fight Club, and the first rule about the Purple Project is you do not talk about that outside of those doors."

Forstall smiled often and spoke confidently as he looked directly into the eyes of the jury. He had testified in several depositions, so he was comfortable on the stand. He came across as a little too practiced to some of the reporters, but they lapped up the information.

During cross-examination, Samsung tried to poke holes in Apple's assertions of the purity of its innovations, introducing emails between Apple's executives that referenced rivals' products in designing the iPhone and iPad. But none was enough to overcome Apple's most damning evidence: an internal email from Samsung's head of mobile communications, J. K. Shin, to his designers in February 2010, right before the Korean company unveiled the Galaxy S.

"All this time we've been paying all our attention to Nokia, and concentrated our efforts on things like Folder, Bar, Slide," Shin wrote, referring to the phone's user interface. "Yet when our UX is compared to the unexpected competitor Apple's iPhone, the difference is truly that of Heaven and Earth. It's a crisis of design."

When Apple's attorney Bill Lee asked Samsung Telecommunications America's mobile strategy chief Justin Denison about the memo, the executive downplayed it, insisting that such exaggerated statements were part of the company's self-critical corporate culture. Samsung didn't want to rest on its laurels and become complacent, he said. "You hear a lot of hyperbolic statements."

"So can you provide documents where Samsung has said the same types of things about Nokia?" Lee returned.

"I'm not sure how I'd do that."

"The answer is that you can't," Lee shot back. "The only mention of 'crisis of design' in all of Samsung's documents is in reference to Apple after the iPhone's introduction in 2007."

In later testimony, Apple landed another blow, introducing another internal Samsung email that mentioned pressure by Google to redesign its Galaxy devices, so they didn't look so much like the iPhone and iPad. "Google is demanding distinguishable design vis-a-vis the iPad for the P3," the email said, referring to the original Galaxy tablet by its code name.

Samsung's defense was that most of the documents comparing the iPhone and Galaxy were just a routine competitive analysis. When

Apple rested its case and it was the other side's turn to summon witnesses, Samsung called one of its senior designers from Seoul to testify that she hadn't copied the distinctive icons on the iPhone's and iPad's home screens. Through an interpreter, Jeeyeun Wang spoke emotionally about how she and her team toiled around the clock to arrive at its own designs.

"I slept perhaps two or three hours a night," Wang said, adding that she had so little time that she hadn't been able to breastfeed her newborn. As she spoke, tears rolled down her cheek.

The designer explained how Samsung's app icons were square like Apple's for functional reasons. "It could not be something that is more of a horizontal type of box or something that's more vertical because to do so would mean that there would not be either enough space or too much space for the finger touching," she said, adding that the image for the phone icon tilted to the left because that was how people made phone calls.

She also explained how Samsung's choice to use a flower to depict the photo gallery app—similar to Apple's icon—was mere coincidence. Wang insisted that they were inspired by a flower wallpaper option in Samsung's televisions.

"When people think of a picture reviewing or viewing a picture, they would think of a landscape that's more or less a horizontal landscape, perhaps a mountain or a river. . . . When we look at close-up shots, generally speaking, people would be thinking in terms of something like a flower."

Samsung also introduced several expert witnesses to try to prove that Apple did not invent the technologies and features such as the rubber band function that it had accused Samsung of copying. It additionally argued against Apple's claim that their products were similar enough to confuse consumers.

The jury listened intently, jotting down notes and occasionally nodding. A few drifted off when the discussion became technical. Several times, Koh paused the proceedings to give the jurors five-minute breaks to stretch their legs.

During the trial, a devastating spoof on Conan O'Brien's late-night show lampooned Samsung as a copycat. The video showed a supposed

Samsung executive as he made a case for how different his company's products were to Apple's.

"Since we entered the personal electronics space, Samsung has created products that, as you can see, bear no similarity to Apple's," he said, showing nearly identical phone and tablet devices. "Notice the grayer edge of our Galaxy phone? And what about our Galaxy tablet? Not even close."

As he changed into an all-black outfit and glasses in the style of the late Steve Jobs, the fake Samsung shill continued. "Samsung's originality is also on display in our home appliances, whether it's our new Macrowave oven, our Vac Pro vacuum cleaner, or iWasher with scroll-wheel controls. Don't believe me? Then come to our retail stores where you can talk more about our products with a Samsung Smart Guy."

Switching to a British accent, the same actor next posed as Samsung's version of Sir Jonathan Ive to explain how the company stayed innovative.

"It's very simple really. We stay true to the vision of Samsung founder Stefan Jobes," he concluded as the screen flashed a mashup of Samsung and Apple's logo with the words "Samsapple—Think Slightly Different."

Samsung was cornered. Faced with the inescapable fact that its smartphone and tablet appeared nearly identical to Apple's, the Korean company's legal team was struggling to offer the jury a coherent counternarrative. During opening arguments, Samsung's lead attorney had acknowledged that his client had been "inspired" by the brilliance of Apple's designs and that this inspiration had spurred Samsung's competitive instincts. Now, despite the most diligent efforts of Samsung's attorneys, evidence was mounting that the inspiration had blossomed into outright thievery. The company's assertion that Apple had copied some of its key mobile technologies inside the phone was buried under Apple's compelling story line.

Samsung's desperation was showing. Again and again, its lawyers were running afoul of Judge Koh. Ahead of the trial, the judge had denied the company's request to show images of iPhone-like designs that were in development before the iPhone as well as a conceptual drawing by one of Apple's industrial designers depicting an iPhone 4–like phone with Sony's logo on the back. Samsung had wanted to use them to prove how unoriginal Apple's designs were, but Koh wouldn't allow it. Her reasoning: Samsung had not previously disclosed or relied on some of the documents to build its case. Some weren't even relevant.

Before the opening arguments, Samsung's lead counsel, John Quinn, had tried one more time to get the slides admitted. "Your Honor, I've been practicing thirty-six years. I've never begged the court like I'm begging the court now to hear argument on this issue," he implored. "What's the point of having a trial? What's the point?"

"Mr. Quinn," Koh responded. "Don't make me sanction you, please!"

Even though he failed to make his case, he had at least established a basis to appeal the verdict later if it were unfavorable. But Samsung wasn't ready to surrender the point. That afternoon, the company disseminated the drawings to the media and took its case public.

"The excluded evidence would have established beyond doubt that Samsung did not copy the iPhone design," said the attached statement, arguing that Apple's designs copied Sony's and were therefore not original. "Fundamental fairness requires that the jury decide the case based on all the evidence."

The judge was furious.

"Call Mr. Quinn. I'd like to see him today," she told Samsung's attorneys after the press release was brought to her attention. "I want to know who released it, who authorized it, who drafted it."

Quinn, however, was at a function with the Academy of Motion Picture Arts and Sciences, for whom he served as general counsel. When the judge was informed of that, she crossly told his colleagues to have him file a report about what happened by the next morning.

Worried that Samsung's stunt had contaminated the trial, Koh questioned the jurors to make sure none of them had seen the news. Apple sought a sanction against Samsung accusing the company and its counsel of trying to prejudice the jury but was denied.

That same morning, Samsung crossed the judge again by quietly bringing in five prospective Samsung witnesses, two interpreters, and three in-house attorneys to see the courtroom without Koh's knowledge. One of them had even asked to take a photograph, which was against the law.

"What's going on here?" Koh demanded. Called onto the carpet again, the lawyers explained that they just wanted to familiarize the witnesses with what a U.S. courtroom looked like. But it was a break from established procedure. Obviously aware that they were pushing the boundaries, the lawyers had sought approval for the visit through a federal judge based in Washington, D.C., rather than through Koh's office.

This infuriated Judge Koh even more. "I will not let any theatrics or any sideshow distract us from what we are here to do, which is to fairly and efficiently try this case."

By the middle of week three, fatigue was settling in. Some of the jurors worried they would have to take more time off from work. Staff attorneys were exhausted after working nearly around the clock with little sleep and uninspired meals of pasta, fish, and beef that were catered into the war rooms. They had been in San Jose for so long that they were running out of menu options. Occasionally, they would sneak out for a proper dinner at places like the Grill on the Alley at the Fairmont, where there would be awkward encounters with the other side.

The judge had everyone on a tight schedule. While the trial was proceeding, teams of attorneys for both companies were preparing exhibits and legal briefs in addition to prepping witnesses and the attorneys in charge. Each evening opposing teams examined each other's evidence and exchanged objections before working through the night on their briefs, which had to be turned in at 8 a.m. sharp. The judge's staff then flew into action, so Koh could hand down decisions on the issues by nine, when court went into session.

Attorneys for Apple and Samsung stressed over whether they had enough time to fully argue their case. Samsung's counsel had devoted so many hours to cross-examining Apple witnesses that they had little time to make their own case. Both sides reminded witnesses that they were "on the clock," asking them to cut to the point. The attorneys frequently sought yes-or-no answers. The testimony wandered through mountains of technicalities. The videotaped depositions of Samsung's Korean executives, which played on a courtroom monitor with voice translations, were mind numbing. At times members of the jury looked ready to dose off, prompting Koh to ask them whether they needed caffeine.

When Samsung threatened to file 140 pages of objections if Apple didn't narrow its witness list, Judge Koh lost it.

"Please don't do this to me. . . . Please . . . I cry uncle," she said, explaining how thinly her staff was stretched to respond quickly to all the motions in a timely manner. "There is just a human limit to what a ragtag team can do compared to your legions of lawyers."

Throughout the trial the judge had repeatedly expressed hope that the two corporations would reach a settlement, thereby saving everyone the trouble of finishing the trial. But toward the end, she conceded that she might be "pathologically optimistic."

As the trial lurched into its fourth week, it was finally time for the jury to deliberate. First, though, the jury had to endure the reading of instructions—109 pages that the two sides had argued over for weeks.

Judge Koh prepared the jury for the ordeal ahead.

"I need everyone to stay conscious during the reading of instructions, including myself," she said. "So we are going to . . . stand up occasionally to make sure the blood is still flowing."

In closing statements, Apple's McElhinny branded Samsung one last time as a copyist. He reminded the jury of all of its evidence showing how Samsung sought to compete against Apple by copying features of the iPhone and iPad. He drew special attention to the fact that no Samsung executives from Korea testified in court in person.

"Samsung has disrespected the process," he said. "Instead of witnesses, they sent you lawyers. Samsung did not call its most important designers and inventors even though we know they were here physically present in San Jose." McElhinny's statement was not entirely fair. Samsung had wanted to call them, but it had run out of time.

Samsung's attorney Verhoeven once again hammered Apple for its anticompetitive behavior. "The real reason Apple is bringing this case is because rather than competing in the marketplace, Apple is seeking a competitive edge through the courtroom," he said, charging that it was "seeking to block its biggest and most serious contender from even attending the game."

Consumers deserved a choice between lots of great products, he argued, telling jurors that their decision could change the future of global competition.

"Is this country going to have vigorous competition between competitors, or is it going to turn into a country with giant conglomerates armed with patent arsenals?" He also made a final case for how the iPhone was evolutionary. "Guess what? Every single smartphone has a rectangular shape, rounded corners, and about ninety percent of the real estate of the front of that phone is the screen," he said. "There's nothing nefarious about this. It's the way technology has evolved. Apple is here seeking two billion dollars in damages from Samsung for alleged ornamentation on that little ten percent around the screen. According to Apple, the way it's interpreting these patents, it's entitled to have a monopoly on a rounded rectangle with a large screen. It's amazing really."

In a rebuttal, Apple's attorney Bill Lee left the jury with one final thought.

"There's a saying among attorneys. If you have the facts, stand by the facts. If you don't, attack your rival's clients, attack their witnesses, and attack their lawyers. And that's what Samsung has done."

In complex cases like this one, it wasn't uncommon for juries to deliberate for days or sometimes weeks. Not this time. That Friday, after just twenty-one hours, the nine jurors informed the bailiff that they had a unanimous decision. The verdict had been reached so swiftly that lawyers on both sides were unprepared. One attorney showed up in a polo shirt and jeans.

"Let me first ask you if the jury has reached a verdict," the judge asked the foreman, Velvin Hogan.

"Yes, your honor."

The twenty-page verdict took a half hour to read. As the clerk announced each decision, the Samsung attorneys looked downcast as they poured over the verdict papers. The company representatives attending the trial also kept their heads down as they jotted notes. A couple of them appeared to be sending the results via the Internet from their seats, presumably to Samsung's Seoul headquarters.

When all was said and done, the verdict was largely a win for Apple, which was awarded more than a billion dollars in damages. Not all of the Galaxy phones and tablets were found to have infringed on every patent, but the jury mostly had found in favor of Apple. Samsung got nothing. The jury determined that none of Samsung's patents had been infringed upon. A Samsung executive in attendance was so upset by the verdict that he angrily brushed past a reporter seeking comment. Apple's stock jumped by $11.73 to $675 in after-hours trading.

Cook was triumphant in an email to employees. "Today was an important day for Apple and for innovators everywhere," he wrote. "Values have won and I hope the whole world listens."

Despite the short turnaround time, the verdict had not come easily. The jury had found Apple's case to be more convincing, especially when they saw the internal Samsung reports and communications that compared the iPhone and the Galaxy. Samsung had also left a poor impression with the videotaped interviews of the Korean executives.

"I thought they were kind of arrogant," juror Manuel Ilagan said. "Maybe it's the culture, but my impression was they didn't want to be there."

The jury had been systematic as they reviewed the instructions that the judge had provided. The verdict form was twenty pages long and complicated, but the group had bonded over the days they had spent together, allowing them to efficiently divvy up responsibilities. Hogan, whom they referred to as "Vel," was unanimously chosen to be the foreman because he had owned patents himself and understood the process the best. Hogan had arrived at least a half hour early on many days during the trial to read the patents and study the evidence. David Dunn, the bicycle shop manager, had organized the various devices for examination. He was comfortable using the devices and could quickly demonstrate whatever function they were examining. AT&T product manager Peter Catherwood polled the group and calculated the damages. Ilagan, a systems engineer, helped explain some of the terminology and the concepts to those who didn't understand them.

Hogan later explained how he had a revelation while watching television after the first day of deliberations.

"I was thinking about patents, and thought, 'If this were my patent, could I defend it?' Once I answered that question as yes, it changed how I looked at things."

Hogan was also adamant that they refrain from addressing Apple's unpatented physical design claims. It was the government's job, not the jury's, to make that call.

"We didn't whiz through this. We took it very seriously," he said. "We didn't just go into a room and start pitching cards into a hat."

The experts' testimony largely fell on deaf ears. The jury concluded that, for hundreds of dollars an hour in pay, the experts were no more than hired guns ready to say whatever their client told them to say. During the testimony, Ilagan had made a note to himself about the experts on both sides: "full of it."

In Munich, the patent expert Florian Mueller was having a restless night when the verdict came in at 1 a.m. Germany time. Upon checking his email on his Galaxy Note phone and finding a news alert from San Jose, he got out of bed to update his blog. Pouring himself a glass of San Pellegrino mineral water, he walked over to his computer in his jogging pants and T-shirt and sat down. After scanning the Twitter feeds and reading the verdict form filed online by the court, he began writing analysis for his blog.

"This ruling is not thermonuclear on its own, but in its aftermath, we will not only see a lot of wrangling . . . but there will also be, even more importantly, a push by Apple to enforce many more design patents and utility (hardware and software) patents against Samsung," Mueller predicted. "There can be no reasonable doubt that Samsung and Google have engaged, and continue to engage, in "copytition" rather than wholly independent creation. Somewhere the courts have to draw the line and afford some degree of protection to the innovators."

For Mueller, the verdict was not big news. He had expected Apple to be handed a large award, but the amount was unlikely to put a dent in Samsung's substantial coffers. The more important question was whether this verdict would give Apple the ammunition to win an injunction against Samsung. What was at stake in the injunctions was ultimately worth many more billions. That battle was still to come.

After responding to a couple of journalists seeking his reaction via email, Mueller sent out one last tweet: "In a few years, the San Jose verdict may—I repeat, MAY—be remembered as the tipping point that sent Android on a downward spiral."

Then he went back to bed.

14

Typhoon

Days after the verdict, Korea came under assault from the worst storm in nearly a decade. Typhoon Bolaven swept onto the peninsula with top winds of 114 miles per hour. Schools closed, flights were canceled, and coastal towns were evacuated. The storm left a trail of at least fifteen deaths and more than $350 million in damages nationwide.

Chairman Lee Kun-hee, however, did not let the typhoon stop him from showing up at Samsung's headquarters that Tuesday at 6:20 a.m. The sky had barely turned light when he stepped out of his black Maybach car. Flanked by bodyguards, Lee entered the lobby of the company's three-building glass complex. Samsung's distinctive buildings stood out among the gleaming skyscrapers. Modeled after puzzle blocks, the design had been chosen to inspire creativity.

Samsung Group's headquarters was located in a forty-four-story building that housed an enormous corporate cafeteria and two day-care centers. That morning, when Lee arrived, the building was virtually empty. The chairman walked past the reception desk and into the elevator. As he rode up to the top floor, a small LCD screen flashed the news of the day.

It had been four days since the verdict across the Pacific. Lee knew the outcome, but his shaken executives had not yet been able to face him to explain how it had gone wrong.

As the storm raged outside in the predawn darkness, Samsung's electronics chief met with the chairman over the abysmal verdict. For two minutes, he was silent.

"Jal hara," he finally said. *Do your best.*

Lee's every syllable was parsed for meaning. This was a man who commanded attention. The same man who had consigned $50 million of inventory to the flames just to make a point to underperforming employees.

He was seventy now, and his black hair was marching toward gray. He had suffered his share of setbacks and humiliations, but he still possessed a ferocious will to dominate. To him, the battle with Apple was part of Samsung's destiny, and he fully expected his company to triumph.

If a killer typhoon didn't slow him down, the old chairman was not likely to be intimidated by the opinion of nine American jurors.

Apple's lawyers had no illusions. As they checked out of the Fairmont and cleared the war room of their files and memos and slides, they were careful not to get too excited. Too much work lay ahead. The California trial was just round one. They still needed to get a court injunction that would block Samsung's products from the U.S. market, and they knew Samsung would keep fighting. Even so, a wave of relief washed over the legal team. All their grueling hours of trial prep—the sleepless nights, the sunrises greeted at their laptops as they drafted yet another brief—had paid off at least for now.

"I can't tell you how much of my life I spent on this," an Apple attorney told an acquaintance.

At Morrison Foerster, the winning lawyers allowed themselves a small victory party, but it was little more than a happy hour with champagne. For many of the staff attorneys, an appreciative note sent around by Bill Lee meant more than any toast. Before the verdict was announced, Lee had written to everyone who had worked on the case, including those outside his own firm. The gist of the email was that, whatever the outcome, they should be proud that they had acted as honestly as possible and had done the best job to represent Apple's interest.

The verdict, however, was already in danger of being outpaced by the realities of the marketplace. What seemed a triumph on the surface was edging toward irrelevance. The phones and tablets that had been part of the case had involved older Samsung models, so the latest products wouldn't be immediately impacted by the decision. Would Apple's win convince more consumers to buy iPhones and iPads? The answer was likely to be no.

Apple's victory had come at a cost that went well beyond the attorneys' fees. When Jobs declared thermonuclear war, he had wanted to paint Samsung as a copycat. Even though Jobs had not lived to see the verdict, he had succeeded. But outside of Apple, few people cared. If anything, Apple's attack on Samsung had validated the Korean company as a worthy rival and supplied it with free advertising. Samsung was already building on its Galaxy successes and using its increased brand power to sell new, original devices such as the smartphone-tablet hybrid Galaxy Note. Samsung had sold more than ten million of the original "phablet" model. Even in the West, it was proving particularly popular among users who appreciated the larger screen but didn't want to carry around both a phone and a tablet.

Around the world, analysts were attempting to divine the true impact of the verdict. Some industry observers worried that Apple's big win could allow big companies to defend themselves more readily and force out smaller competitors. In the smartphone industry alone, an estimated $20 billion had been spent on patent litigation and patent purchases between 2010 and 2012. Apple and Samsung had each spent tens of millions of dollars in legal fees.

"When patent lawyers become rock stars," said former Apple general counsel Nancy Heinen, "it's a bad sign for where an industry is heading."

Apple's win also put an uncomfortable spotlight on its motivations. Why was the company wasting so much time, money, and energy protecting its older technologies if it had game-changing products up its sleeve? Could it be that there was nothing more in the pipeline? Steve Jobs was famous for never looking back. But perhaps the company now had too much to lose.

UBS analyst Steve Milunovich posited that the victory could come back to haunt Apple. "The real threat is not a competitor beating Apple at its own game but instead changing the game," he wrote in a research note. "The likelihood of Apple being leapfrogged or a rival creating a new category is greater if they have to think outside the box."

So far, the innovation that Apple had demonstrated since Jobs's passing had been lackluster. Siri so far held great promise but its underperformance compared to Apple's overhyped marketing had sabotaged its credibility. A year after the service's hapless debut, Siri was a tangential feature at best.

Apple had been down before—in a hole much deeper than this one—only to claw its way back, ultimately proving itself more resilient and profitable than anyone could have predicted. But if the company was going to get back on track, it needed another dazzling advance—another innovation that would restore its reputation as a game-changer.

Apple was plotting another coup, this time by air. For months, a fleet of planes and helicopters had been flying around the world—above the Sydney Opera House, Big Ben in London, the Transamerica Pyramid in San Francisco, as well as numerous other locations from Europe to Asia to the United States. Operating at low altitudes, each aircraft was equipped with cameras that captured the topography below.

The empire was mapping the globe. For years, Google Maps had been a default app on iPhones and iPads. Now Apple was quietly working to supplant that feature with a mapping app of its own. In the battle for mobile dominance, control over the feature was crucial. If Apple could weave its own mapping technology more deeply into its devices and improve on the user's experience, that advance could become a key point of differentiation. With a more sophisticated mapping function, for example, Apple's calendar could inform users when they needed to build in more traveling time to account for heavier traffic.

Google was becoming an increasingly powerful competitor, and Apple was reluctant to rely so heavily on the enemy. Tension between the two sides had mounted in 2008 during discussions to renew their agreement over the mapping app. Apple wanted features that Google offered in its Android version, such as turn-by-turn navigation and the ability to see a photo of a location from the street. Apple was also concerned about the data that Google was gathering from the app. Location information was becoming very valuable as marketers sought to target more relevant ads to consumers. By giving it away to Google, Apple was making its competitor stronger.

The breaking point came as Apple worked on an update for its iOS mobile operating system for the iPhone 5. Though more than a year remained in the contract between the two companies, Apple decided to replace Google's map app with its own. In anticipation, the company acquired several firms, including a 3-D mapping company called C3 Technologies, based in Sweden.

Apple's plan was to introduce the new application at WWDC that summer of 2012, but blogs and news media spilled the news ahead of time, depriving Apple of the big reveal. The first hint had come the year before when researchers discovered that iPhones were collecting location data and transmitting them back to Apple. The company's acquisitions of mapping companies and the growing strain between Apple and Google had been reported as well.

By that June, when WWDC opened, any mystery surrounding the unveiling of the maps had been shattered—a bad omen, given Apple's need to protect its glittering reputation. After Siri, the company needed to get this right. Cook and his team had to impress the audience with a flash of the old magic.

The moment the keynote began, though, it was clear that something was off. In a wince-inducing introduction, the company chose to open with Siri telling cheesy jokes.

"Hello and welcome to WWDC. I'm Siri, your virtual assistant," said the familiar robotic voice, emanating from an iPhone shown on a big screen. "I was asked to warm up the crowd, which should be easy since the high will be seventy-five degrees."

A set of drums played a rimshot. *Ba-dum-tshh.*

"So here we are in San Fran, the ATM of Silicon Valley. If you developers need investors to finance your app, I found three hundred ninety-six venture capital firms fairly close to you."

The drums played again. *Ba-dum-tshh.*

Soon, Siri was taking cracks at Apple's rivals.

"Hey, any of you guys working with ice cream sandwich or jelly bean," Siri asked, naming Google's names for its Android operating system. "Who's making up these code names. Ben and Jerry?"

Ba-dum-tshh.

"But seriously, I am excited about the new Samsung. Not the phone. The refrigerator. Hubba hubba."

For Apple, which always banked on its cool, the opening was jarring. In years past, the company had held itself above the competition. Now it was using Siri to make cheap potshots. The gag was painful, because it hinted at a growing desperation inside the empire. Besides, anyone with an iPhone knew that Siri was incapable of banter—even lame banter. Apple insisted on pretending otherwise.

The strained enthusiasm was still showing when Scott Forstall bounded onto the stage.

"All riiight . . . let's talk about iOS," he began. "I find it *incredible* that through the end of March we have already sold more than three hundred and sixty-five *million* iOS devices."

Forstall spread his arms wide to illustrate the magnitude of the accomplishment. In talking up the demand for iPhones and iPads, he appeared to be giving himself a big pat on the back. As the lead manager of the mobile operating system, Forstall took credit for much of the iPhone and iPad's success. Revenues from mobile devices now eclipsed that of Apple's traditional mainstay Macintosh business.

Forstall eventually turned to the maps app.

"We have built an entire new mapping solution from the ground up, and it is beautiful," Forstall said, boasting about how the company had done its own cartography. He showed how users could easily look up information about businesses and see where there were traffic jams. Forstall was particularly proud of the 3-D mapping feature, which Apple had dubbed "Flyover."

"We have been flying major metropolitan areas around the world," he said, demonstrating how you could pull up the Sydney Opera House and rotate it to see the cityscape in the background. "It is gorgeous!"

The overenthusiasm, the parade of forced superlatives, the silly put-downs of the competition—all of it seemed to betray a deeper insecurity.

Apple's growing defensiveness was even more apparent three months later when the company unveiled its latest phone. A small rotating pedestal rose from the stage, and a tiny spotlight beamed onto the iPhone 5.

"It is an absolute jewel," Schiller gushed. "It is the most beautiful product we have ever made, bar none."

As reporters craned to see, he showed it on the screen so they could get a better look.

"It is really easy to make a new product that's bigger. Everyone does that."

The audience understood that he was taking a jab at the Galaxy Note. Again, the remark signaled Apple's increasing vulnerability. If Samsung was such a lousy company with such inferior products, why was Apple mentioning them at all? Schiller just kept going.

"That's not the challenge," he said. "The challenge is to make it better and smaller."

Sticking to the script, the marketing chief gave a loving description of the iPhone 5—how its case was made entirely of glass and aluminum, how at only 7.6 millimeters it was the world's thinnest smartphone, 18 percent thinner than the 4S. It weighed just 112 grams, which was 20 percent lighter. Its screen size was four inches, up from three inches, allowing for an extra row of apps. The pixel count in the display was 1,136 by 640. Its width was just big enough for most thumbs to maneuver things on the screen while holding it in the same hand. Its length allowed users to see more content. The processor was twice as fast. It was compatible with LTE wireless networks, which were ten times as fast as the older 3G networks. As always, Schiller radiated enthusiasm, but his breathless litany of specs made him sound like a stereo salesman pimping another amp.

Then came a video of Jonathan Ive, talking about the creation of the new phone as though it were a holy relic.

"Never before have we built a product with this extraordinary level of fit and finish." Ive explained how complex its manufacturing process was. The surfaces of the aluminum enclosures were machined before being polished and textured. The symmetrical sloping beveled edge of those frames, called the chamfer, was cut with a crystalline diamond to create "a near mirror finish."

Despite the executives' best efforts to jazz up the launch, the presentation was long on technical features and notably short of the sense of mystery that Jobs had so powerfully conjured. A Japanese newspaper reporter who used to write about Sony was struck by how Apple's presentations now resembled the product launches of other electronics makers, crowing about details that few consumers cared about. The wonder was gone.

Apple had never been about technical specifications. It was late to integrate 4G networks, and it had yet to offer near field communications, an emerging technology that promised to allow users to easily exchange data or make purchases by putting devices near each other. Consumers bought Apple products because they seemed magical. While its rivals tried to one-up each other with the thinnest, lightest, or highest-resolution phone, Apple focused on the look, feel, and user experience, which was usually captured with a bewitching tagline designed to evoke an immediate emotional response in consumers. The original iPod: a thousand songs in your pocket. The original iPhone: the Internet in your pocket. The iPad: a magical and revolutionary device.

The iPhone 5's tagline: "The biggest thing to happen to iPhone since iPhone." Unless Apple was literally talking about the larger screen, it was unclear from the presentation what the big deal was. The tagline was uninspired.

Focusing on technical features invited unwanted comparisons to its competition. Several of them, including Samsung, manufactured many of the key technologies themselves. Reviewers would later line up the iPhone 5 with Samsung's newest model. Though they were positive overall, they noted how the iPhone 5 lacked the long battery life and memory of the Galaxy SIII. Most people didn't understand or care about the subtleties. Everything that they wanted already existed in the phones they were carrying, so these latest improvements seemed incremental and boring.

That quarter ending in September, Apple's smartphone market share fell to 14.6 percent from a peak of 23 percent at the beginning of the year. By contrast, Samsung's share rose to 31.3 percent, more than triple its share from just two years earlier. Though analysts had initially expressed concern that Apple's screen suppliers would be unable to keep up with demand, the opposite turned out to be true. Demand was so much lower that Apple was soon forced to cut its order in half.

The luster was fading.

Apple could not afford to make another blunder. But in the days after the iPhone's launch, the howls began anew.

Anticipation had been building for the new mapping application, pre-installed in the iPhone 5. But once it was released, the reaction was disastrous. Forstall and his team had been so intent on beautifying the mapping system that they had overlooked the fact that the app itself was unusable: missing roads, misplacing labels for businesses and landmarks, and riddled with errors. Even more so than Siri, Apple Maps became the subject of ridicule as consumers and bloggers took to Facebook, Twitter, and other sites to complain or mock mistakes. In searches, Apple's app labeled the city of Berlin as "Schöeneiche bei Berlin," referred travelers to an airport in Dublin that didn't exist, and brought up London, Ontario, instead of its better-known U.K. counterpart. Around Washington, D.C., the map couldn't find Dulles Airport or the Kennedy Center. The water between Asia and Australia was mislabeled as the Arctic Ocean. Greenland sank under the waves and became the Indian Ocean. An Apple executive was a half hour late to a meeting because the application had guided him to the wrong place. The errors were so bad that the humor magazine *Mad* put Apple Maps on its list of the "Top 20 Dumbest People, Events and Things of 2012."

The level of inaccuracy was startling, particularly given that mapping technology was well established. Reliable databases of mapping coordinates were readily available, and other companies had already solved many of the major problems. Apple's inability to make it work was bewildering.

"It's quite hard to mess up that bad on such fundamental layers," said Marcus Thielking, a former executive at German navigation software maker Navigon who helped run a mapping start-up called Skobbler, which built its app using some of the same data as Apple. He suspected that some of the issues occurred when Apple tried to improve performance by reducing data size or tweaking the amount of data the app consumed. If Apple was using rough GPS signals from users' iPhones to help build its map, that could also create problems since the phone location data can be noisy and unreliable. To offset that, most mapping app makers went through a process called street matching, in which they matched the signals to the map they had. So if people were moving in a grid where there was nothing, they would investigate to see if there was a street there that they had missed.

Apple may not have encountered this many problems had it stuck to a simpler app. But Forstall and his team were so eager to supplant Google Maps that they may have become too ambitious. Developing such a complex mobile application was extraordinarily difficult. The process required battalions of people, lots of coordination, and exhaustive testing, especially if you were taking data from multiple sources that then needed to be matched up and aligned with where things were in real life. Google had legions of staff working on maps around the world. Smaller start-ups used an army of volunteers who fanned out in any new location to run extensive accuracy tests for months before debut. At Apple, Forstall had fewer than one hundred people working on maps. As with Siri, the project had been developed quietly, and Apple had kept the team purposefully small. The company had failed to run enough tests to assure reliability.

Once again, Apple's arrogance and secrecy had trumped common sense. It had received advanced warning of the problems and yet had gone ahead with the launch anyway. Developers who had been testing the maps application had filed bug reports and complained. One developer who received a response was told that the issues were "well understood." But little appeared to have been done to fix them. The question was why? Who knew about the extent of the problems but thought it was okay to proceed? It was unlikely that Forstall hadn't known, but did Forstall keep the problems from Cook, did Cook know and decide to proceed anyway, or did Cook give the green light without making sure the application was ready? Here was another indication of Apple's engineering process breaking down.

The complaints began hours after its release. By the next day, as the cries grew into a storm of disbelief and outrage, Apple's leadership was already in damage control mode. Cook wanted Forstall to issue a mea culpa in his name, but the iOS chief balked. Phoning in remotely to a meeting with other executives, he told Cook that Apple should handle the maps problem the same way that it had dealt with the iPhone 4 antenna issues—no apology. If there was to be a letter to the public, he thought it should come from Apple rather than himself.

Jobs had encouraged such dissent because he thought such vigorous debates helped the company reach the best decision, but Cook hated public posturing.

Two days after the maps app launched, Cook issued a personal apology.

To our customers,

At Apple, we strive to make world-class products that deliver the best experience possible to our customers. With the launch of our new Maps last week, we fell short on this commitment. We are extremely sorry for the frustration this has caused our customers and we are doing everything we can to make Maps better.

We launched Maps initially with the first version of iOS. As time progressed, we wanted to provide our customers with even better Maps including features such as turn-by-turn directions, voice integration, Flyover and vector-based maps. In order to do this, we had to create a new version of Maps from the ground up.

There are already more than 100 million iOS devices using the new Apple Maps, with more and more joining us every day. In just over a week, iOS users with the new Maps have already searched for nearly half a billion locations. The more our customers use our Maps the better it will get and we greatly appreciate all of the feedback we have received from you.

While we're improving Maps, you can try alternatives by downloading map apps from the App Store like Bing, MapQuest and Waze, or use Google or Nokia maps by going to their websites and creating an icon on your home screen to their web app.

Everything we do at Apple is aimed at making our products the best in the world. We know that you expect that from us, and we will keep working non-stop until Maps lives up to the same incredibly high standard.

TIM COOK
Apple's CEO

In an unprecedented move, Apple directed users to the competition's apps. Though Cook tried to sound an upbeat note about how many customers were already using Apple Maps, this was an indirect admission of how deep the problems were. They were unlikely to be resolved anytime soon.

Apple rarely showed such deep contrition. Jobs had usually opted for defiance even when he was obviously in the wrong. During the brouhaha over the iPhone 4 antenna, much of the discussion about public relations strategy had revolved around how Apple could address the issue without "your tail between your legs," as Jobs's longtime PR guru Regis McKenna put it. But that attitude had fed a growing image of Apple as arrogant. Cook wanted to change that image and viewed this as an opportunity to make that shift.

Cook never forgave Forstall for refusing to apologize. A month later he asked his mobile software chief to leave the company.

News of Forstall's departure fell like a bombshell. Even the most senior members of the iOS team learned of his firing just moments before it was publicly announced. Though the mapping catastrophe was the trigger, there were broader reasons for his departure. As a colleague in Jobs's management team, Cook must have seen Forstall's political maneuverings over the years as he took credit for accomplishments but blamed others for hiccups. Forstall was the least liked executive on the team. Since Cook took over, he had also been complaining, saying that there was no "decider" anymore. Without Jobs, Forstall didn't have a protector.

The question in many people's minds, both inside and outside the company, was what his departure meant for Apple. Aside from Ive, Forstall was thought to most embody Jobs's vision and spirit of innovation. He was a smart, capable software engineer, and Apple couldn't afford to lose such talent just because he didn't always work well with others. Some considered him to be an eventual CEO.

But if Cook couldn't corral Forstall and direct his abilities in a productive way, he had to let him ago. "Collaboration is essential for innovation," Cook said in an interview. "We're brought together by values. We want to do the right thing. We want to be honest and straightforward. We admit when we're wrong and have the courage to change. And there can't be politics. I despise politics. There is no room for it in a company."

Cook gave iTunes head Eddy Cue responsibility over Siri and Maps and his Mac software chief Craig Federighi control over iOS. Both were team players. The addition of Siri and Maps to the iTunes and iCloud team also made logical sense. All of the cloud-based services were now folded under one umbrella.

Jonathan Ive took control over the team that designed the way people interacted with their devices. In many ways the decision made sense. Under Forstall, Apple had been pursuing what was known as skeuomorphism, a kind of visual design that maintains the ornamental characteristics of the old physical tools they stood in for. The calendar app, for example, had the kind of faux leather stitching that a desk calendar might have, and the books in iBooks were made to look as if they were sitting on wood shelves. The use of textures and shadows was pervasive in Apple's software. Though it was Jobs who initially encouraged this style of design, the consensus lately had been that Forstall's team had taken it too far. The podcast app referenced a reel-to-reel tape deck. Game Center was dressed in lacquered wood and green felt to make it resemble a casino. These elements were now deemed irrelevant and tacky, and they needed to be more aligned with the elegance and understatement of the devices themselves.

No one was seemingly better equipped to update the look and feel of the software than the person responsible for the physical design of Apple's products. Designing people's interactions with Apple's devices was a natural extension of what Ive already did. The lead designer had lobbied to control this group for years, but Jobs had never allowed it. Cook was more willing to delegate responsibilities.

John Browett was another one of his casualties. Since the retail executive joined Apple in April 2012, Browett had made sweeping changes to the Apple Stores. To boost profits, he had cut costs—laying off some store staff and cutting the work hours of those he had kept. He also put a greater focus on meeting sales targets and incorporated Dixons-like sensibilities into the stores' layout by demanding that more accessories be displayed on shelves. In a company where the store experience was built around making technology accessible and where staff had been trained to provide Ritz-Carlton class service regardless of whether they made a sale, Browett's actions were wildly unpopular. The staffing cuts in particular created such an uproar that the executive was forced to admit to his staff, "We messed up."

Just six months after his arrival, Browett was fired. The announcement of his departure was little more than a tersely worded footnote in the press release about the broader management changes.

"I just didn't fit within the way they ran the business," Browett later said. "For me, it was one of those shocking things where you are rejected from the organization for fit rather than competency."

Fourteen months after Cook took over, he had finally shaken up the leadership team. But it had taken several missteps. Siri's problems may not have been Cook's fault, but how had he allowed the same pattern to repeat itself with maps, which fell squarely under his watch?

And why had Cook picked Browett to run retail in the first place? The strategies Browett had tried to implement at Apple had been consistent with his experience at Dixons and Tesco. The culture at those retailers was oceans apart from Apple's. Why Cook thought he would be a great fit had been a mystery to many.

By this time, Cook's management style was becoming more apparent. Cook delegated responsibilities and rewarded his executives as long as they did well. But if they made a mistake, he came down on them hard. The danger with that approach was people becoming risk-averse and stifling innovation.

All of these considerations raised the question: Was Cook the best choice to chart Apple's future?

Apple, as it happened, was devoting considerable energy to studying the dynamics of leadership.

Before Jobs died, he had created Apple University, an internal think-tank and management training center.

Joel Podolny, the former dean of the Yale School of Management, headed the center. Another key member was Richard Tedlow, an esteemed Harvard professor who had studied the careers of some of history's greatest business innovators such as Andrew Carnegie and Henry Ford.

Podolny's hand-picked team had free rein to talk to any of the executives, and they built case studies such as how the company had created its retail strategy. The curriculum included examples from the outside, ranging from the collapse and subsequent bankruptcy of the A&P grocery chain to the design process for Manhattan's Central Park. Occasionally they invited guest lecturers to talk to Apple's executives.

One of them was a Harvard professor, Gautam Mukunda. A political scientist-turned-management expert, Mukunda had studied the role of a leader in organizations under pressure. He arrived in Cupertino as Cook was grappling with the Apple Maps fiasco.

Mukunda—a former student of Clayton Christensen—had formulated a theory that leaders could rarely be categorized simplistically as "good" or "bad." What was more important was whether they were "filtered" or not. Filtered leaders were people who had risen through the ranks, were entrenched in the culture, and were chosen after a careful internal process of selection and evaluation. In contrast, unfiltered or extreme leaders were often volatile characters who slipped through and got the job because of a particular set of unusual circumstances. Because they were outliers, they tended to do things differently. Their tenure was either a brilliant success or a catastrophic failure, but rarely in between.

Filtered leaders were surer bets, but they were also relatively interchangeable because they more or less rose through the same process. Unfiltered leaders were riskier. While most failed, sometimes the payoff was huge.

Jobs wasn't a typical unfiltered leader because the company he led was his own creation. But the fact that he spent years in exile had made him an unlikely CEO candidate for Apple. He would never have been given a second shot if Apple hadn't gone nearly bankrupt. Saving the company had required someone who understood the culture but could think in fresh ways—someone with the nerve to take risks.

Cook was a classic example of the filtered leader, not just at Apple but also with respect to corporate America. Cook had started his career at an iconic American company, IBM, and obtained an MBA before spending fifteen years at Apple, studying every move that Jobs had made. Cook had been fully indoctrinated into Apple's way of seeing the world. He had also mastered the mundane logistics. He did not pretend to be an innovator or a visionary. His gift allowed him to see inside the mysteries of the supply chain, not the future.

Apple's two CEOs were exemplars of Mukunda's theories. The professor believed that Jobs had made the right decision in choosing Cook as his successor instead of turning to another wild card. Unfiltered leaders were a huge gamble. Given the high rate of failure of such leaders, the chances were low that Apple would have found two spectacular unfiltered successes in a row. Cook might not have been able to perform the kind of amazing feats that his predecessor had made look so easy, but his organizational skills allowed him to build on the advantages that Jobs had created. More importantly, Cook could transition the company so it revolved around a system rather than an individual.

With a filtered leader at the helm, Mukunda believed Apple would inevitably evolve into a different company. In an industry where there were few obvious products left to reimagine, Cook was unlikely to orchestrate more game-changing ones. He wasn't wired for risks. The question was, would that kind of steady, less innovative business be an acceptable outcome for Cook, the board, and Apple's shareholders?

Cook had often repeated Jobs's advice to not dwell on the past. At the same time, Cook had also stressed that Apple's culture would not change.

"Apple is this unique company, unique culture that you can't replicate. I'm not going to witness or permit the slow undoing of it because I believe in it so deeply," he had told investors at the Goldman Sachs conference in February 2012. "Steve drilled in all of us over many years that the company should revolve around great products. And that we should stay extremely focused on a few things, rather than try to do so many that we did nothing well. And that we should only go into markets where we can make a significant contribution to society."

In the face of increasing competition from Android in the mobile market, however, Apple's culture might need to change for the company to continue thriving. Cook needed to show more willingness to depart from Apple's past.

To Mukunda, it was a simple matter of what he called business physics.

"There are just forces in any environment and any market that constantly drag companies to the mean," said Mukunda, referring to factors like complacency or margin pressure by competitors. "What Apple did was essentially in violation of business physics for an extremely long time. They created this beautifully optimized machine."

The problem with that strategy was that, when consumers stopped caring as much about the things Apple was optimized for such as design, they might choose to focus on something else that didn't play to the company's strengths.

The fundamental question, Mukunda believed, was what kind of leader did Cook want to be? If Cook wanted to focus on profits and consolidate Apple's leadership in the industry, then he just needed to put the right people in place to churn out the next Macs, iPhones, and iPads. If he wanted Apple to stay a revolutionary company, then he needed to bring in new, unfiltered leaders who were hungry and bold.

"Apple can be an excellent ordinary company or a genuinely extraordinary one," Mukunda said. "But it can't be both."

Either way, Cook needed an executive team built around his strengths, weaknesses, and vision. His current team was still largely Jobs's team, assembled around Jobs's personality. The likelihood that they could be successful under someone else was low.

"The team that's got them where they are," Mukunda said, "is not the team that's going to get them where they're going to go."

The company was arguably facing its biggest threat since Jobs had brought the company back from the brink. When Apple unveiled the iPhone and then the iPad, it had shown swaths of rivals the way to what Jobs often talked about as the post-PC era, in which traditional computers would be supplanted by smartphones, tablets, and other connected non-PC devices. Now everyone was chasing the same vision. Most obviously there was Google. But Amazon.com was an ambitious company with a strong following and the resources to compete against Apple. Facebook understood how people liked to connect better than anyone else. Apple's old rival Microsoft was still very much in the game with its Windows Phone operating system. Companies like Research In Motion and Sony hadn't given up, either. Many of them were led by unfiltered leaders. Amazon.com had Jeff Bezos. Facebook had Mark Zuckerberg. And Google had Larry Page. Meanwhile, Apple had Cook, the self-proclaimed "Attila the Hun of Inventory."

Around the world, people still remembered Steve Jobs. But the mention of Tim Cook drew blank stares among ordinary consumers.

"Who's that?"

At most companies, becoming a household name was irrelevant to a CEO's performance. But Apple presented a special case. For all their acumen Cook and his team were still stuck in the genius trap. A year after they had taken the reins, their company was still overwhelmingly defined by a cult built around a dead man. If Apple was ever going to truly move forward, its leadership had to replace that cult with a startling new invention, or another creative force. So far, Cook had shown himself to be either unwilling or incapable of filling that role. He was an unreadable mystery.

The most recognizable personality at Apple was Siri. Hapless. Confused. Devoid of soul.

Revolt

Two days after the iPhone 5 went on sale, a riot erupted in Taiyuan, in northern China.

The incident underlined the deeply interconnected nature of the global economy. In California, Apple ordered millions of new phones to meet preorders and initial sales. In China, Foxconn received the order and instructed its managers to rev the factory lines. The factory managers turned to the supervisors who ran the production lines, telling them to lean harder on their crews. The pressure, already surreal, suddenly spiked. And the workers, who had had enough, revolted. Up until then a few of them had sought release by jumping off buildings. Now, inside Foxconn's factory, they turned their rage outward.

Gangs of employees—some estimates put the total number as high as two thousand—tore gates off hinges, broke windows, and damaged cars. Riot police were sent in to quell the violence. Dozens of people were hospitalized. Production stopped for a day.

Sitting in their offices in Cupertino, Apple's executives had no way of knowing that this latest order would push the supply chain past its breaking point. All they knew was that the iPhone had a new design for the first time in two years, that their target audience was growing, and that their projections suggested that this phone would break all sales records. They couldn't look into the hearts and minds of the hundreds of thousands of young men and women toiling on the other side of the planet to make those projections come true. All they had was their numbers, so orderly, so clean on their laptops.

Foxconn officials attributed the unrest to a personal dispute that spun out of control. But workers blamed the conflict on security guards who had severely beaten a man in a mini-bus after a fight started in the dormitory. When other workers from the same province learned what had transpired, they became angry. In the pressure-cooker environment, the tension ignited. More workers joined the riot. The roughly two hundred guards who were on duty were soon overwhelmed.

"The guards here use gangster style to manage," one of the workers, Fang Zhongyang, told a reporter outside the campus gates. "We are not against following rules but you have to tell us why. They won't explain things and we feel we cannot communicate with them."

In the aftermath, security teams with helmets and plastic shields patrolled the grounds. A looped recording of a voice asking workers to maintain order blared over a loudspeaker as the factory resumed production. Guards at the entrance were on high alert. The slightest disturbance was quickly quelled. Guards reprimanded workers who chatted too loudly while waiting to enter the grounds. They also yelled at workers seen talking to reporters.

"Stop talking!"

"Be quick!"

Before Apple and Foxconn had a chance to recover from the bad news, another incident broke out, this time at Foxconn's iPhone 5 factories in Zhengzhou, in north-central China. Workers and quality control inspectors went on strike over what they perceived to be unreasonably high production standards and inadequate training.

Apple was always demanding about quality, but manufacturing this latest model was extra challenging because of its design. The backs of the previous two models—the iPhone 4 and 4S—had been glass with a stainless steel frame. But this time, both the back panel and the edge were made out of the same aluminum that was used in its notebook computers. The designers liked the material because it looked sleek and was considerably lighter weight than glass and steel. The problem was that aluminum was malleable and led to frequent scratching and chipping.

Foxconn was expected to somehow overcome that shortcoming. The impossible task rolled down from the managers to the quality control inspectors to the line workers. To keep the production lines going, many of the line workers were asked to give up Golden Week, a seven-day holiday kicked off with National Day, a celebration commemorating the founding of their country. The pressure came to a head in early October.

The details of what transpired next in Zhengzhou are unclear. According to China Labor Watch, the New York–based advocacy organization that first reported the strike, Apple had instructed Foxconn to raise its quality control standards after receiving complaints from customers about scratches on the iPhone. When the inspectors stepped up their scrutiny of the production line and began rejecting products, some of the workers revolted and beat several of them. In frustration and anger, the inspectors had then initiated the worker strike.

"Workers are under tremendous pressure," wrote Wang Sheng, a factory worker, in an online chat on the Chinese microblogging site Sina Weibo. "Foxconn and the clients still set rigid quality standards. We have to bear all the difficulties in production." Wang used the pseudonym Ye Fudao. When his real name became public, he stopped communicating with reporters.

Both China Labor Watch and Wang estimated that three thousand to four thousand workers had gone on strike, and multiple iPhone 5 production lines were paralyzed for an entire day. But a government spokesman in Zhengzhou told China's official Xinhua News Agency that the strike had involved only about one hundred inspectors, who had refused to work for an hour after one of them was beaten by workers.

Other local media, who interviewed Foxconn employees, estimated that the workers who directly participated in the strike had numbered in the hundreds. One publication reported that the strike lasted for two days, not one.

Foxconn itself gave at least two different explanations. In a statement, its Taiwan-based offices insisted, "Any reports that there has been an employee strike are inaccurate."

A press release said the incident involved two brief and small disputes. "There has been no workplace stoppage in that facility or any other Foxconn facility and production has continued on schedule." The statement added that workers had been paid triple time to work through the holidays.

One of Foxconn's spokespeople separately told the Chinese financial newspaper *Securities Daily* that about four hundred workers were "absent from work" for two hours.

Wherever the truth lay, two such controversies in such a short time span was damaging to both Apple and Foxconn. After the earlier suicide and labor scandals, this latest turmoil again reminded the public of the darker side of how their beautiful new iPhones and iPads were made.

The workers' riot and the strike exposed ominous cracks that threatened the future of Apple and Foxconn's symbiotic relationship.

A certain degree of labor unrest was inevitable for a massive manufacturer that had grown to employ more than a million workers in a society undergoing major socioeconomic changes. In a communist country that prohibited independent labor unions, workers' only recourse to express their dissatisfaction was through spontaneous strikes or protests. Foxconn wasn't unique. It was just singled out in the media because of its association with Apple. Few Asian manufacturers could withstand the kind of scrutiny that Foxconn received.

But if Apple brought Foxconn unwanted attention, the opposite could be said as well. Two partners that had worked so well together for nearly fifteen years were increasingly becoming liabilities to each other. The growing problems between Apple and Foxconn were particularly evident in the sprawling complex of factories where the strike had reportedly broken out. The fissures threatened Apple's supply chain—and its future.

Until recently, the massive Zhengzhou campus had embodied the culmination of Foxconn's wildly successful partnership with Apple. Terry Gou, Foxconn's dynamic CEO, had built the Zhengzhou facility to manufacture tens of millions of iPhones, and his ability to deliver on those ambitions was a testament to the tremendous influence that Foxconn now wielded in China. Henan, the province where Zhengzhou was located, had long been notorious for its thieves and petty criminals. Gou had won the governor's support by promising to remake the province.

"Everyone said I cannot change Henan, where there is only stealing and deception without any high-tech background," Gou told the governor. "It's okay, I will make the change for you by bringing you the finest, the iPhone!"

Close to three hundred thousand employees worked on the secluded grounds, which encompassed nearly two square miles. It was Foxconn's biggest location in China. The land had once been a sleepy agricultural site known as "the Village of Jujubes." The oldest jujube tree was said to be more than one thousand years old, dating back to the Song Dynasty. In record time, the government relocated the trees and the farmers who tended them, built apartment buildings for workers, and established "The Xinzheng Comprehensive Tariff Free Zone." Zhengzhou's politicians had seen how Foxconn's presence in Shenzhen had transformed that province into the leading economy in the nation, elevating its government leaders to prominent positions in the Communist Party. They hoped that the new factories would provide Henan with the same economic boom and that their careers would prosper, too.

Foxconn built its campus in one hundred days. The complex opened in 2010, and within two years was churning out two hundred thousand iPhone 5s daily. Farmers put down their shovels to sign up for jobs at the factory. The base salary Foxconn offered—1,800 yuan a month—may have been insufficient for workers in Shenzhen, but in Zhengzhou, it was about 70 percent higher than the local minimum wage. With overtime, a farmer could make in two to three months what he used to make in a year.

The properties around Foxconn's complex soon turned into prime real estate for shops, entertainment venues, and restaurants that catered to the factory workers. With a major airport just ten minutes away and a new high-speed rail system that connected it to Beijing five hundred miles to the north, Henan Province was poised to grow into an important transportation hub: The Chicago of China, so to speak.

Though Foxconn made products for almost all of the top electronics brands in the world, Apple's business mattered most. When the two companies first began working together, neither had anything to lose. Back in 1998, Apple was a has-been close to bankruptcy, weighed down by huge inventory buildups in its warehouses and factories. Foxconn was a small company with little experience. Apple was able to close its factories in the United States and make products more cheaply without compromising quality.

As Apple soared, so did Foxconn. In just seven years, Foxconn jumped 328 spots in *Fortune* magazine's Fortune 500 list, moving from number 371 in 2005 to number 43 in 2012. Its 2012 revenues reached more than $130 billion—nearly a third of Taiwan's gross domestic product. Analysts estimated that about 40 percent of its revenues came from Apple. Its profits surpassed the combined total from its rivals—Wistron, Quanta, Inventec, Pegatron, and Compal.

But if the two companies had ascended together, they also risked faltering together. Their intertwined fortunes were underscored by Apple's disappointing earnings report for the quarter ended September 2012. Helped by the launch of the iPhone 5, the company had sold more phones than analysts had predicted, but iPad sales had missed expectations. Concerns about Apple's performance spilled over to Foxconn's business even though the manufacturer's earnings report was strong that quarter.

After doubling and tripling revenues and profits for several years, Apple's business growth slowed to 20 to 30 percent. At other companies, in other industries, that growth rate would have been more than respectable, but the drop was steep enough to bring the negatives of Apple's business to the forefront and upset the dynamic between the two companies. Once Apple's business slowed, the very premise of their mutually beneficial liaison began to break down. That made Apple vulnerable.

Over the years, Terry Gou had agreed to every demand that Apple made. Gou had acquiesced to decrees on costs, quality, and deadlines. To keep up with the production orders from across the Pacific, Foxconn had trained hundreds of thousands of new workers and had invested in factories in Zhenghzou and elsewhere. As Foxconn's association with Apple brought greater scrutiny, the company opened its doors to journalists and tried to be more transparent about its business. When questions about labor conditions surfaced, Foxconn raised wages, curbed overtime, and addressed other complaints.

In years past, even when Apple's operating margin topped 30 percent, Gou had been willing to live with as little as 1.5 percent. Even with that differential, Apple's business was crucial to his company's future.

Like Apple, Foxconn was at a crossroads. Nearly forty years after Gou had founded the company, the business was maturing. Already manufacturing electronics goods for practically every major brand in the world, the company now found itself with little room left to expand. To keep growing it, Gou needed to transform Foxconn and reduce its reliance on Apple. And it had to happen soon. Gou was already in his early sixties, and he had pledged to retire at seventy.

Virtually unknown in the United States, Gou was a towering figure in Asia, every bit as ambitious and charismatic as Steve Jobs. Unlike many business owners in Taiwan who came from privileged backgrounds, Gou was a self-made man.

After starting his company in the mid-seventies, he had boldly pursued business in the United States largely on guts and willpower. Gou had traveled across America in the early 1980s, visiting companies in thirty-two states in eleven months. Though he was on a shoestring budget, he had rented a Lincoln Town Car in every city so he could make a stylish impression. In anticipation of his future success, he practiced signing his name in English until he perfected it.

His big break came at IBM in Raleigh, North Carolina, possibly in the same offices where Cook had worked. Desperate to secure an order for connectors, Gou camped out in the lobby for three days until he got an appointment. When Compaq was looking for a partner in Asia, Gou won the contract over bigger rivals with his determination and a belief in unbeatable customer service.

"There was nobody like Terry," remembered Greg Petsch, a Compaq operations executive. "It was like, 'You need me to expand my factory, I'll put the cost in there and won't charge you if you don't fill my capacity.'"

Gou put out the red carpet on Petsch's first trip to Taiwan. When the plane door opened, there stood Gou with a customs official to escort his customer off the plane before anyone else. After bypassing customs through a special exit, a car stood waiting to speed Petsch and Gou away. With other contract manufacturers, the CEOs might make an appearance once and then let their underlings handle the relationship, but Gou was front and center every time Petsch flew in. Foxconn's CEO provided five-star service. When it came to pleasing potential business partners, no demand was too great.

When Cook moved to Apple and hired Foxconn to make its computers, Gou approached the business with the same relentless drive. His relationship with Apple's executives was so close that when he married his second wife, both Cook and Jonathan Ive attended his wedding banquet in Taipei. Apple executives referred to him as "Uncle Terry."

Inside his own company, Gou was terrifying. Behind the conciliatory veneer was a general who required his managers to squeeze profits out of the toughest contract requirements. Internal meetings with Gou were known to be as harrowing as meetings with Cook. Gou would curse and interrogate his subordinates, and he openly scolded people to make an example of them. Unlike Cook, Gou was a bit of a showman and preferred to talk rather than listen.

"Gou cannot talk to an empty meeting room," a former Foxconn executive told a Taiwanese magazine. "There has to be many people sitting there, and he will not talk but curse at them to get his inspiration."

In one meeting, he had supposedly called on one of his executives. "What are the eight screens?" he demanded. "Name them in the correct order."

"Cell phone, tablet, notebook, all-in-one, portable TV, TV, electronic billboard, LED big screen."

Gou barked back, "Why did you name them in that order? Is it by size? Wrong! It's about eye distance."

Gou's point was that the quality and type of screen needed in a product was not determined by the size of the device but by the distance that the user was looking at it. A touchscreen, for example, doesn't make sense in a sixty-inch television screen that is viewed from a long distance. His men were awed by his ability to think differently.

Gou had a reputation in the industry for being a megalomaniac. Among his top lieutenants, an allegory was told about Gou ordering an executive to knock down a load-bearing wall. The executive nodded, but didn't follow through. A few days later, Gou asked again, "I told you to knock it down last time. Why didn't you do it?"

"Yes, yes, yes," the executive responded as he pretended to obey the order by picking up a hammer. The third time Gou passed by, he threatened punishment, so the executive knocked down a tiny bit of the wall. When Gou discovered that the wall was still there, he was livid, so the man finally obeyed.

But when the walls of the building predictably came crashing down, Gou told the executive, "So if I told you to jump off the building, will you really jump off?"

Governed by a founder-CEO with that kind of total control, Foxconn faced the same leadership challenges as Apple, especially as Gou looked for someone to succeed him. Except Gou didn't have a Tim Cook. He was said to start his day by personally approving travel requests from his one hundred direct reports. Many of his senior executives were too old to take the reins of the company, but the younger managers were not yet capable of such enormous responsibility. Even if Gou hoped to keep the company in the family, his two eldest children by his first wife had so far expressed no interest in running the business. One worked in the film industry and the other in the financial sector. A daughter and son by his second wife were still preschoolers.

According to one former Foxconn advisor, succession was a particularly difficult challenge in Asian companies because the culture encouraged compliance and uniformity. Executives tried too often to fit in, not stand out.

"What I worry, really worry," said the former advisor, "is whether they have the ability to create new things beyond contract manufacturing."

Gou needed to redirect Hon Hai before he retired. He had to find a way to minimize the company's exposure to the ebbs and flows of Apple's business.

Some analysts and industry insiders had speculated for some time that Gou could eventually move into other more profitable industries beyond contract manufacturing. The CEO had already expressed a strong interest in mobile applications and cloud-computing technology and was hiring thousands of software engineers. He spent $840 million to acquire a 37.6 percent stake in Sharp's biggest LCD panel factory and began manufacturing flat-screen televisions that he then sold through partners like RadioShack in China, Vizio in the United States, and Chunghwa Telecom in Taiwan. The company had aspirations of eventually providing content for all of the devices it assembled. Separately, Foxconn was said to be considering selling its own brand of accessories, including those that were compatible with iPhones and iPads. In late 2012, the company acquired a $200 million stake in the high definition portable camera brand GoPro.

As part of a vision to become a one-stop shop, Foxconn was also revamping its distribution services and retail operations in Asia. At the other end of the supply chain, Gou was investing in highly profitable component businesses like display maker Chimei Innolux.

Some observers believed that Gou secretly hoped to establish his own electronics brand. Gou's denials did not end the speculation. For years he had served Apple and other high-tech houses, watching those companies hoard the king's share of the profits as they tossed him crumbs. If he had decided to keep those billions in profits pouring from his factories and his workers, who could have blamed him?

Whatever move Gou made next, the days of Apple taking his company for granted were waning. Apple's near-total reliance on his factories and his cheap labor put the California company in a position of weakness. Over the years Foxconn had made itself more and more indispensable by embedding itself into its partner's operations. After manufacturing so many millions of iPhones and iPads, Gou's company had absorbed and internalized Apple's perfectionism and idiosyncrasies. It was difficult for any of Foxconn's rivals to mass-produce those devices at the scale and quality and speed Apple demanded. As the balance of power shifted, Foxconn gained the leverage to push back against Apple and negotiate higher prices. When Gou's team was allowed the discretion to choose parts suppliers, the company gave the business to its own subsidiaries to add to its gains. Working with business partners dictated by Apple, Foxconn pushed aggressively for rebates.

Tim Cook had once touted the necessity of being aggressive and unreasonable with its suppliers around the world. Was it any surprise that Foxconn had learned that lesson so well?

The real problem, though, went much deeper.

Apple's supply chain was no longer under Apple's complete control.

Apple had no factories of its own. Most belonged to Foxconn.

Apple had no factory labor force of its own. Foxconn had a million workers willing to work for almost nothing as they mounted production lines twenty-four hours a day.

Apple was losing leverage to dictate. Foxconn was less willing to accept its terms.

Uncle Terry was gone. To an unsettling degree, Apple's future was now in the hands of General Terry.

Apple recognized the danger and was trying to reduce its dependence on Foxconn. The company stepped up its efforts to spread its business among more manufacturers. According to *DigiTimes*, a Taiwanese industry publication owned by a businessman with close ties to contract manufacturers, Apple planned for a Taiwanese company called Pegatron to eventually make 60–70 percent of its iPads and a small portion of its iPhones.

In addition to Pegatron, Apple had been working with one of Foxconn's archrivals—BYD, a well-known Chinese manufacturer of cars and rechargeable batteries. Apple began by ordering BYD batteries for its phones, but its hope was that the Chinese company would take on more and more of the manufacturing. The speculation in the industry was that Apple intended to turn them into another major final assembler.

To ensure that no company held the kind of sway that Foxconn did, Apple also started parceling out its business in smaller quantities.

Replacing longtime suppliers wasn't easy. Apple owned the intellectual property for its devices, but the production expertise belonged to its suppliers. At the end of the relationship, the Foxconns and Pegatrons of the world took that proficiency and put it to work for other customers while Apple had to train another supplier.

A case in point was Apple's breakup with AmTran, a Taiwanese company that made monitors for desktops. Several years earlier, Apple decided to end its contract with AmTran and move the job to another Taiwanese company, Quanta, which already made its all-in-one iMacs. Soon, though, Apple realized that Quanta's workers could not maintain the same level of quality. Apple tried to solve the problem by asking AmTran to share with its competitors the know-how it had acquired in making the monitors. But when Apple sought to bring a team from Quanta into AmTran's production facilities, they were denied entry. Neither threats nor pleadings worked. They even offered to wear blindfolds, so they didn't see other clients' products that were being made in the same facilities. The answer was still no.

Apple's justification was that the intellectual property embodied in the monitors was theirs, so they should be allowed to bring in anyone they wanted. AmTran rejected the argument, citing confidentiality reasons. In the end, Apple was forced to work with AmTran for many more months.

This behavior—trying to force cooperation from a partner it had unceremoniously dumped—only reinforced Apple's reputation for hubris. Foxconn was far from alone in its grievances against the boorish Americans from California. Apple's bad reputation was so widespread that it threatened relationships with many suppliers.

Horror stories about working with Apple abounded in the industry. A familiar pattern soon emerged. Apple seduced suppliers with the promise of a big contract. When the companies opened up their businesses, Apple probed every detail from their technology expertise to their finances and cost structure. Next, auditing teams descended on the suppliers, sweeping through the factories to inspect infrastructure, production capacity, workforce size, and other details. The auditors often included people who had previously worked in the industry. Later, the procurement team would leverage that information to wrangle better prices and terms for Apple.

The price negotiations never stopped. Under Jobs they had taken place about once a quarter, but under Cook they occurred as frequently as every month. Partners weren't even allowed to benefit from the association because they were sworn to secrecy and forbidden from publicizing Apple as a customer. Apple's faithful operations corps had become as ruthless and relentless as Cook had trained them to be.

Even after a deal had been struck, Apple kept close tabs on production, requiring daily quota projections as far as a month or two out. Operations team members camped out at suppliers' facilities to make sure the quotas were met. If something went wrong, some of them would get so angry that it would frighten some of the workers. Once, after an accident at Pegatron, an Apple manager from the United States slammed a notepad on a desk in frustration. The workers who witnessed the scene talked about it for days. Some, including Foxconn's corporate managers, found the cursing so intolerable that they quit.

Getting on Apple's preferred supplier list was no guarantee of business because Apple could suddenly switch to a cheaper supplier or a different technology. In Taiwan, one of the recent cautionary tales was of touch panel screen maker TPK. For a long time, the company built screens for iPhones and iPads, turning itself into a global leader. In 2011, TPK's expertise with Apple's products allowed sales to double to more than $4.9 billion from the previous year. But then Apple switched to a thinner screen made by companies such as Sharp Electronics, LG Display, and Toshiba Mobile Display. TPK saw its shares plunge.

In Japan, even worse stories circulated. The most shocking tale came from a company called Shicoh, which blamed Apple for driving it into bankruptcy. The small company, located in the outskirts of Tokyo, made tiny motors to help focus smartphone cameras. Apple's operations team had asked the company to build a new clean room and invest in new equipment in preparation for work on the iPhone 4S. The company, already a preferred Apple supplier, took that as a sure sign of a deal and obliged, even seeking additional cash through a public offering to pay for the improvements. But shortly thereafter Apple suddenly cut them off because their financial statements were not strong enough to meet Apple's criteria. The business collapsed. Apple took the contract to a rival Japanese company, Alpine, which had been secretly building a cutting-edge, automated manufacturing facility for Apple, even as Apple negotiated with Shicoh.

The potential rewards of working with Apple had once justified putting up with such atrocious tactics. But as sales slowed and the company began divvying up its business into smaller chunks, companies began to wonder if the time had come to send the bully packing.

"Until about a year ago, the market didn't pay attention to us unless we were an Apple supplier, but now they see it as a risk," one Japanese electronics parts supplier told *Diamond Weekly*, a Japanese business magazine that conducted a three-month investigation into Apple's suppliers in Japan. The reportage, which listed twenty-seven partner companies, flagged the extreme dependency of Japan's technology industry on Apple's business.

"There is no clear answer for whether suppliers should accept the risk and go after Apple's business, or pull out once they find themselves overly dependent," the article concluded. "But Apple's presence has become too big for Japanese manufacturers. How they deal with this giant in the years to come will determine which suppliers live or die."

A similar sentiment was felt in Taiwan. "The benefits at present don't diminish the fact that Apple has always eaten the meat while the suppliers have only shared the soup," said Lan Guanming at *CTimes*, an industry trade monthly. He asserted that Taiwanese suppliers should continue to take good care of Apple, but they should also keep an eye out for better customers.

In a sign of Apple's diminishing influence, one executive of a former supplier agreed to meet with a journalist at a hotel lobby in Taipei in the fashionable Xinyi district near the landmark Taipei 101 building. During Apple's glory days, when Jobs was still CEO, this executive would likely have never spoken to a reporter.

On a Monday morning, the man showed up dressed in a dark suit despite the sweltering weather outside. After settling down in the lounge with coffee, he discussed the terms of the interview—not overtly, as was typical in the West, but in a veiled way that still provided the assurance that he and his company would remain anonymous.

Once the conversation began, he spoke more frankly. He explained that Apple categorized its suppliers into three tiers. First-tier partners like Foxconn had more leverage, but the secondary or tertiary suppliers had to be willing to take on a big risk for a potentially small return.

"It's better to serve eighty percent of the market," he said, adding that Apple had an impossibly high standard for quality and interfered too much in their business. "They dictate who can go in and out of their factories."

In a refrain that was repeated in the supplier industry across Asia, he also found the foul language that the Apple managers used to be offensive. The rewards of the business were not worth the indignities and the stress.

In the last weeks of 2012, as Apple's fortunes showed further signs of fraying, the risks borne by the suppliers mushroomed. Apple's market share growth in the smartphone market was slowing, while Samsung's share was soaring, helped in part by an aggressive marketing campaign that cost more than $400 million in the United States alone. Making matters worse, the demand for the iPhone 5 was clearly not meeting projections. The suppliers' nightmare came true when Apple informed its LCD panel makers that it was slashing its orders by more than half. Already deeply troubled by its struggling television business and fighting for survival, Sharp in particular was devastated. The Japanese display company had an entire plant dedicated to Apple that it had to let go idle.

Sharp was desperate. The upkeep for the factory cost $100 million per month, a loss they couldn't afford. When the company's executives asked Apple if they could start working on the panels for the next-generation iPhone instead, the answer was no. Not only that, but Apple demanded that Sharp spin off the dedicated iPhone factory to shield those assets from the rest of the business in the event of a bankruptcy. "We didn't realize the strong side effects of doing business exclusively with Apple," a Sharp executive lamented. "It's like we bit into a poisoned apple."

Until then, the company had been firmly in Apple's camp; in addition to making iPhone screens, the company shared ownership of an LCD screen factory with Foxconn. Nevertheless, in March 2013, Sharp accepted an offer from Samsung to buy a 3 percent stake in the company for $111 million. The company ignored Gou's fierce objections and agreed to supply Samsung with screens.

Orders from Apple recovered within a few months as production for the next iPhone model ramped up, but all the maneuvering now placed Sharp in a delicate situation. The company had potentially put itself in the crossfire of three global powerhouses.

Foxconn wasn't immune to Apple's slowing sales. That spring, the company stopped hiring at its factories and postponed plans to expand its Zhengzhou plant. In the first six months of 2013, its shares plunged nearly 20 percent as investors and analysts worried about where Foxconn's growth would come from.

At an eight-hour shareholders' meeting in June, Gou unveiled plans to strengthen his business by spinning off undervalued divisions and beefing up areas like research and development, software, and intellectual property rights. He also apologized for his company's poor performance.

"Please give me some time," he said. "I work sixteen hours every day and I also work on weekends to plan for the company's future growth. I won't let you down. The good day is not far away."

From Apple's perspective, Gou's statement begged an important question: Where did Apple fit in the General's plans?

Velvin

The jury foreman would not stop talking. In the weeks after the Apple-Samsung verdict, Velvin Hogan granted interviews to at least five news organizations, dishing on how the jury had reached its billion-dollar verdict.

Hogan told Reuters that he and the other jurors had wanted to make sure that the damages were painful to Samsung but not unreasonable. He told the *San Jose Mercury News* that the verdict was intended as a billboard-sized warning to corporate copycats. To the *Verge*, a technology blog, he proclaimed that serving on the jury had been a great honor—the high point of his career.

"You might even say my life," he added.

In the aftermath of most high-profile trials, jurors are reluctant to speak publicly, preferring to return to the quiet of their lives and let the verdict speak for itself. But Hogan welcomed the spotlight, going so far as to join a combative online discussion with Gizmodo readers, patiently fielding their written questions for hours, as well as enduring their insults and innuendos. Many of the participants, clearly outraged by the verdict, suggested that there was no way the jury could have understood the complexities of the case. They took Hogan and his fellow jurors to task for not seizing the opportunity to nullify troublesome aspects of patent law. One called the verdict an "epic fail." Why hadn't the jurors made Apple pay for trying to patent a rectangular-shaped phone with rounded corners? Had they simply ignored Samsung's arguments? How could they possibly have read the judge's instructions and paid attention to the evidence? Had they even understood the significance of the case? Although the participants steered clear of the word *bribe*, several pressed Hogan on how much money Apple had paid the jury.

"Not a dime," answered Hogan.

The foreman maintained his composure. He pointed out that it wasn't the jury's job to rule on the legitimacy of patent law. He and the other jurors had sworn to follow the law, he said, and they had been faithful to that oath. If Gizmodo's readers believed the legal system was flawed, he said, then they should work to change the law. His unwavering calm enraged his audience. Noting that Hogan himself worked as an electronics engineer and owned a patent, they wondered how he could claim to be fair and impartial. Wasn't he passing himself off as an authority? And wasn't that inherently problematic, especially when arriving at the verdict in a case of such magnitude?

"Why don't you try to be honest," wrote another, "and tell us how your ego really feels?"

Digging for any pro-Apple bias, several asked if Hogan owned an iPhone. No, he said, pointing out that none of the jurors had owned any Apple devices. This revelation aroused more suspicion. Wasn't that, someone pointed out, a statistical anomaly? How had the case possibly ended up in the hands of a Silicon Valley jury devoid of iPhones?

"I just can't believe much of what you're saying," wrote one skeptic. "I think Apple found a way to compensate the jury somehow, but I may be wrong there. However, believing that none [of] the jurors own an Apple product is too much."

The grilling went on and on. When one of Hogan's responses contained a misspelled word, his inquisitors pounced.

"If you can't speak proper English," one wrote, "maybe you weren't the best choice. . . ."

Even by the dismal standards of online comments, the Gizmodo discussion was appalling. For all their harping about the apocalyptic implications of Hogan's typos, many of his critics couldn't string together a grammatically correct paragraph. Cloaked in anonymity, they were free to indulge in not just cruelty but aggressive hypocrisy. None of them appeared to have attended the trial or heard the testimony and evidence. No one showed any particular command of patent law.

The real question was why Hogan had subjected himself to a public whipping. Had he not foreseen how nasty it would get? Did he view the online forum as a chance to defend himself and the rest of the jury? Maybe the foreman simply saw it as his duty to explain the reasoning beyond a billion-dollar verdict. Many of his online comments displayed a touching, almost naïve belief in the primacy of good intentions. He explained that he and his fellow jurors had been well aware of the importance of the case, that they had listened to the evidence, followed the law, and rendered what they believed to be a just decision. As he put it:

"I/we the jury stand by our ruling."

Hours went by, and the mob grew more livid. They painted Hogan as a fool and a dupe and a troll, and compared him and the other jurors to Nazi war criminals, and asked him how it felt to be so wrong, and did he prefer crack or more traditional forms of cocaine, and was he simply pro-Apple or also anti-Samsung, and how did he sleep at night, and what kind of name was Velvin, anyway? They also vented their wrath toward Apple, ranting about the company's arrogance, accusing it of being the true thief of innovation, and despairing over how everyone kissed its corporate ass. They wondered if the jury's verdict was fueled by an anti-Korean, pro-American slant and asked Hogan several more times how much Apple had paid him for the verdict, and how much Apple was paying him now to deny that he had ever been paid. They brushed aside his denials and told him, come on, did he truly expect anyone to believe he wasn't on the take, and if Apple could patent a geometric shape, did that mean it could patent an orange and sue anyone who sold oranges, and did Hogan realize he was a nobody, and a terrible person, and did he understand that every word he uttered might destroy the verdict?

By then, the foreman had gone silent. But it was too late.

Venomous as the free-for-all on Gizmodo had been, the questions about Hogan's impartiality were not so easily dismissed. After dissecting his many statements from the media circuit, Samsung's legal team filed a motion alleging juror misconduct and requesting a new trial. The lawyers noted that the foreman had revealed several misunderstandings about patent law. According to his answers in interviews,

Hogan erroneously believed that design patents were based on the look and feel of a device, and that a device had to be entirely different to be cleared of infringing a utility patent. Other comments he'd made, the lawyers argued, could be seen as proof of bias in favor of patent holders such as Apple.

"In this country," Hogan had told Bloomberg Television, "intellectual property deserves to be protected."

The most damning accusation involved an incident from Hogan's distant past that he had not disclosed during jury selection. In his post-trial interviews, Hogan mentioned that decades ago he had worked for Seagate Technology, a company now partly owned by Samsung. After Hogan left, Seagate had sued him for breach of contract, ultimately forcing him to declare bankruptcy. In their motion for the new trial, Samsung's lawyers argued that the foreman had deliberately omitted any mention of the Seagate dispute—and its connection to Samsung—because he wanted to serve on the jury.

"Mr. Hogan's failure to disclose the Seagate suit raises issues of bias that Samsung should have been allowed to explore in questioning," the Korean company said in its filing.

Hogan defended his lack of disclosure, saying that he was only obligated to tell the court about litigation that he had been involved in over the past decade.

"Had I been asked an open-ended question with no time constraint, of course I would've disclosed that," he told Bloomberg. "I'm willing to go in front of the judge to tell her that I had no intention of being on this jury, let alone withholding anything that would've allowed me to be excused."

Apple called the two-decade-old lawsuit between Hogan and Seagate "irrelevant." If Samsung had a problem with it, counsel should have raised the questions during jury selection. The controversy continued to swirl as the media pointed out that there had been no time frame stipulated when Hogan was asked whether he had been involved in a lawsuit.

Skeletons from Hogan's past were only one of several issues raised with the court. Apple may have won a billion-dollar verdict, but Samsung was still fighting to have the damages reduced or thrown out.

More important, the crucial battle over the injunctions had not been resolved. The key question was whether Apple could get an injunction on all of the twenty-six Samsung devices that were found to have infringed its patents.

Apple's request for an injunction wasn't heard until December 6, three and a half months after the verdict. That afternoon, attorneys for both sides reconvened at the San Jose courthouse. Reporters found themselves once again in the back of the courtroom. The agenda included Samsung's appeal in addition to the injunction.

For the first couple of hours, the lawyers sparred over the damages. Back and forth the two sides battled, filling the courtroom with a fog of words. Apple's lawyers talked about "the mootness claim." Samsung's lawyers insisted that Apple's arguments were "completely fanciful," "astonishing," "preposterous." Somehow a billion-dollar argument descended into tedium. When the attorneys suggested they might need to file more paperwork, Judge Koh expressed frustration. Whichever way she ruled, she knew an appeal was inevitable. Couldn't they just get on with it?

"I was hoping," said the judge, "to send you on your merry way to the federal circuit."

When the discussion turned to Apple's injunction request, Koh wanted to know: If Apple was no longer using an old iPhone or an iPad design, then how would Samsung's use of a similar design hurt Apple?

Apple's attorney drew a comparison to the automobile industry. If Chevrolet came out with a car that looked like a 1967 Ford Mustang, would that be okay if Ford wasn't making that car anymore? In a similar way, allowing Samsung to sell products that looked like the iPhone 3G or 3GS would dilute the uniqueness of Apple's designs.

By the time the iPhone maker had made its case, three hours had gone by.

"I'd like to finish before 2013," the visibly fatigued judge said as she gave Samsung an opportunity to respond.

The company's counsel argued that an injunction would result in hardship because the ban would create fear among retailers about selling Galaxy devices.

As the hearing drew to a close, the exasperated judge asked, "When is this case going to resolve? Is there some endpoint here?"

The room broke into laughter. But Koh wasn't kidding.

"For Samsung, these are dollars-and-cents decisions," said Apple's attorney Harold McElhinny, accusing Samsung of calculating its moves based on whether they thought the courts could catch up to them. "They make the decision every day how close they are going to get to the line."

He described the damages awarded as "a slap on the wrist." Apple was looking for more than that. The courts needed to stop Samsung from copying the company's products going forward.

"I'll be frank. If it ends up that we get a new trial on damages and there's no injunction and there's no—if you didn't see the same case that the jury saw, then I'm not sure how we get to a resolution from there," McElhinny said.

Samsung's attorney Charles Verhoeven argued again that Apple was trying to eliminate its competition through the courts rather than the marketplace. Samsung was willing to negotiate, he said, but the ball was in Apple's court.

McElhinny protested that Apple had agreed to a meeting of their executives, but the talks had never happened because of Samsung.

The judge encouraged the two sides to find some way to find resolution.

"It's time for global peace," she said. "It would be good for the consumers. It would be good for the industry."

In the last ten minutes of the hearing, John Quinn, the Samsung attorney who had gotten into trouble with the judge during the trial, spoke for the first time to broach the issue of jury misconduct. Velvin Hogan, claimed Quinn, had been deliberately dishonest.

"Your Honor, what do we know about the foreman in this case?" the lawyer asked. "This is a juror that just wanted to be on the jury."

When Koh asked why Samsung hadn't inquired about the relationship with Seagate during jury selection, Quinn said they would have asked, but they didn't know about the dispute. He asked that Hogan and the other jurors be brought back into court for questioning.

"Your Honor," said Quinn. "It would be an abuse of the court's discretion at this point not to hold a hearing."

One of Apple's attorneys rose to speak.

"Very briefly," Koh warned.

William Lee denied that Hogan had deceived anyone. "It's outrageous that he's being called a liar. They're claiming that . . . it was his goal in life to get on the jury."

The attorney pointed out that Hogan's legal troubles with Seagate had occurred in 1993 and that Samsung had not acquired a 10 percent holding in Seagate until 2011, only several months before the trial. Why would the foreman have harbored any ill feelings about Samsung?

"The most preposterous part, your honor, is this: They're claiming Mr. Hogan lied about an event that occurred nineteen years ago, that he harbored a grudge for nineteen years," said Lee. "That doesn't make any sense."

The judge responded noncommittally. She thanked the lawyers and promised to issue her rulings as soon as she could.

Eleven days later, Koh denied Samsung's request for a new trial. She dismissed the suggestion of jury misconduct, saying that Samsung had failed to prove whether Hogan had intentionally concealed his lawsuit or engaged in any misconduct. She also said that Samsung had been given ample time to question him during the trial about his employment at Seagate, which Hogan had mentioned in the jury selection process.

"What changed between Samsung's initial decision not to pursue questioning or investigation of Mr. Hogan, and Samsung's later decision to investigate was simple," the judge wrote in her order. "The jury found against Samsung and made a very large damage award."

But the judge also denied Apple's request for an injunction banning Samsung's devices. Her reasoning was that it was unnecessary. Samsung no longer sold most of the phones in question. The company also had already changed their software to avoid infringing on Apple's patents.

The ruling was a blow for Apple. What the judge said was true— the products that would have been banned were obsolete. But the injunction would have given Apple ammunition to pressure Samsung into making changes throughout their newer lineup. Courts around the world were watching closely, so a favorable decision by Koh had the potential to tilt the scale for Apple in other battles, too.

The $1 billion hardly made a dent in Samsung's coffers. At the end of 2012, Samsung's electronics business alone had about $35 billion in cash. The war had never been about money. The real battle was over market share and protecting the iPhone and iPad's competitive advantage.

The verdict had at least allowed Apple to reach a ten-year licensing agreement with HTC in November. Though the terms were confidential, experts believed that the Taiwanese smartphone maker felt the pressure to reach a deal because HTC had been accused of violating some of the same kinds of patents as Samsung. Unlike the Korean company, HTC couldn't afford the prolonged battle or the potential damages if they lost their case.

A settlement at some point between Apple and Samsung seemed inevitable as well. But on what terms?

After arguing in front of Koh about who was to blame for a face-to-face meeting that wasn't held, the two sides met the following week in Seoul, where they agreed to another meeting in January. Proposals and counterproposals were sent back and forth across the ocean, but nothing had come of the talks.

Samsung had no reason to settle. Apple had yet to land a knockout punch against Samsung anywhere in the world. The iPhone maker had bagged some early wins in Europe and Australia, where courts banned Samsung from selling some of its devices. But some of the injunctions were later scaled back or overturned. In other cases Samsung had quickly developed workarounds to avoid violating Apple's patents. In yet another case between Apple and Samsung that was proceeding around the same time as the San Jose trial, a judge at the U.S. International Trade Commission ruled in favor of Apple on four patents. The decision potentially had immense ramifications because the ITC had the authority to ban sales or imports to the United States, but the decision carried no weight until it was approved by the full commission in a process that would likely take months.

Even in the California case, there was no material impact on Samsung until the appeals court ruled. The company didn't have to pay damages yet, and the immediate danger of an injunction had been averted in December.

As Samsung's executives had promised Chairman Lee, the company was fighting back hard. In the fall of 2012, courts in Tokyo and the Netherlands handed Samsung wins, declaring that the Korean maker did not infringe on Apple's patents. Meanwhile, an anonymous entity—possibly Samsung—was submitting reexamination requests to the U.S. Patent and Trademark Office on some of Apple's patents. In response, the office had invalidated Apple's "rubber band" patent, which Samsung had been found to have infringed in the California trial. The patent would remain valid through the appeals process, but the decision threw an additional element of uncertainty into Apple's case. Samsung's strategy was clear. It was planting seeds of doubt in the public about the strength of Apple's claims.

All the while, Samsung was continuing to arm itself with even more mobile patents that it could use as ammunition against Apple. By the end of 2012, Samsung had become the largest mobile patent holder, topping industry pioneers like Nokia, Ericsson, Alcatel-Lucent, and Qualcomm. Apple didn't even make the top ten.

A few months later, Judge Koh handed Samsung another advantage. She struck $450.5 million from the $1 billion in damages that Apple was awarded because the jury had incorrectly calculated them. A new trial would have to be held to determine the damages on fourteen Samsung devices. Apple could potentially be awarded even higher damages than before, but either way, Apple's ability to claim its win was postponed yet again. Samsung had been remarkably effective at pushing the narrative that the jury had no idea what it was doing and Apple didn't deserve the magnitude of the win that it scored.

The failure—at least until then—to protect its innovations was a bitter pill for Apple. When the company developed the iPhone, Steve Jobs had patented everything. He had trusted that the legal system would enforce those patents. But up against a formidable, wealthy opponent like Samsung, Apple's innovations seemed poorly protected even after millions of dollars in legal and expert fees as well as countless hours of executives' time.

Apple appealed Koh's denial of the injunction, but a decision would take months.

Jobs had wanted to go thermonuclear, and under Cook's leadership Apple had followed through on that threat. But to what end? The lawyers were still filing their motions, arguing back and forth, and Samsung was still making billions in the United States on its phones and tablets. At the end of 2012, Samsung's global smartphone market share was nearly 40 percent. Apple's was 25 percent. A year earlier, the two companies had been neck-and-neck.

By participating in the trial, Apple had also given up many development secrets. None may have been material information, but the reveals diminished the company's mystique.

Even if Apple had been able to win an injunction, the impact of it was debatable. Samsung had devices in the works that steered clear of Apple's patents. The assumption was that they would be inferior, but that wouldn't necessarily be true.

Samsung had lost a verdict. But Jobs's strategy wasn't winning the war, either.

Critical Mass

Outside the courts, Apple basked in the spotlight. The company was considered the pride of America—a reputation confirmed in February 2013, when President Obama lauded the company during his annual State of the Union address.

Tim Cook had been invited to the ceremony as one of Michelle Obama's guests of honor. The tradition had started in 1982 when President Reagan had invited a man named Lenny Skutnik, who had helped save a passenger from a jetliner that had crashed into the Potomac River. Since then the invitees, from Rosa Parks to Sammy Sosa, were known as Skutniks. Though some considered the tradition to be political theater, the guests embodied what each president thought America stood for.

As a champion of the technology sector, Obama had included people from the industry before. Last year, Jobs's widow, Laurene, had been a guest along with Instagram cofounder Mike Krieger. Obama had spoken about how America should support risk takers and entrepreneurs who aspired to be the next Steve Jobs. This year Cook had been invited along with two dozen other guests of honor, including a teenage inventor who had devised a low-cost way of detecting pancreatic cancer; a first-grade teacher from Newtown, Connecticut's Sandy Hook Elementary School, site of a mass murder in November 2012; and a lesbian couple fighting for equal rights for same-sex military couples.

Sitting behind the first lady, Cook looked distinguished in a conservative black suit and tie. Earlier, Jack Andraka, the teenage inventor, had asked to take a photo of him. Andraka had posted it online with the message, "taking a picture of Mr. Cook w an iPhone! priceless!"

Midway through his speech, President Obama addressed the country's manufacturing industry. "After shedding jobs for more than ten years, our manufacturers have added about five hundred thousand jobs over the past three. Caterpillar is bringing jobs back from Japan. Ford is bringing jobs back from Mexico. After locating plants in other countries like China, Intel is opening its most advanced plant right here at home," he said. "And this year, Apple will start making Macs in America again."

As the president spoke, he looked up at Cook. Cameras captured the moment as they zoomed in on the Apple CEO, past Michelle Obama, who looked dazzling in her black and maroon Jason Wu dress. As Cook smiled, he emanated an aura of establishment.

A former Apple executive watching the speech on television was appalled. Nowhere was there a trace of the subversive upstart that Jobs had founded. Though Jobs had relationships with government leaders, and Al Gore served on his board, the executive was sure that Jobs never would have attended. He didn't have time for such pomp and circumstance.

Having the president of the United States make the announcement in such a high-profile manner was nonetheless a coup for Apple, especially considering that the move was mostly a token public relations gesture. Cook had already told the media that he was planning to bring some of the manufacturing back.

"We've been working on this for a long time, and we were getting closer to it," he had told *BloombergBusinessweek* in an interview the previous December. "We could have quickly maybe done just assembly but it's broader because we wanted to do something more substantial."

The reality was less grand. Apple was only investing $100 million, a pittance compared to its $137 billion cash hoard. The company wasn't even planning to assemble the Macs themselves. The media believed Foxconn would be handling the manufacturing in the United States because the Chinese manufacturer had made suggestive comments about making more products in America. In actuality, Apple had reportedly given the business to Singapore's Flextronics. It was telling that the job was too insignificant for Foxconn to care.

Despite the president's stamp of approval, the truth was that Apple's star was falling. Scarcely more than a year after the death of Steve Jobs, his beloved creation was beset by so many challenges, emanating from so many fronts. It wasn't just the dearth of dazzling new game-changing devices, which made so many wonder if Apple had lost its magic, or the embarrassments that had turned the iconic company into a late-night punch line. It wasn't just the harsh attention that had been cast on the company's aggression, its mania for control, and its struggles to redefine itself. Or the dangers inherent in Apple's reliance on Foxconn's factories and massive workforce. Or the reality, so clear now after Judge Koh's denial of an injunction, that Apple's war against Samsung was at an impasse, and Android was winning the war for dominance in the marketplace.

Any one of these setbacks would have been enough to send tremors through a company that sought to reinvent the world. But it was all of this, coalescing at the same time, created a wave of momentum, pulling Apple down.

The bad news kept piling up. A few months after Apple's disastrous launch of its own maps app, Google showed up its rival with a superior new app for the iPhone. Google's new app featured turn-by-turn navigation and schedules for more than one million public transportation stops around the world. Within forty-eight hours of its release, the app was downloaded ten million times. While the iPhone's functionality was improved by the app, headlines proclaiming Google's stunning achievement only highlighted Apple's navigational failures.

In mid-March, a U.S. district judge in Manhattan ordered Tim Cook to sit for a four-hour deposition in the e-book antitrust case that the Justice Department was still pursuing. Arguing for Cook's deposition, the government attorneys said the CEO was likely to have relevant information about Apple's entry into the e-book market, including conversations with Jobs. But Apple had been fighting the motion, saying that whatever Cook could share would only repeat the testimony of eleven other Apple execs who had already been deposed. The judge agreed with the Justice Department, saying that the impossibility of getting testimony from the now deceased Jobs required others to step in. The trial was set for June.

On March 15, 2013, a day after the judge ordered Cook's deposition, Apple was thrust back into damage control mode when a crisis erupted on the other side of the world. China Central Television, a government-controlled network, was accusing Apple of showing disrespect to Chinese customers—a serious allegation in a market of great importance to Apple's future.

In a glitzy two-hour prime-time show honoring World Consumer Rights Day, the network targeted Apple in an investigative report that accused the company of treating Chinese consumers as second-class citizens. Specifically, the program said Apple skirted Chinese laws by offering just one year's warranty rather than two as was required. The show also claimed that when customers experienced problems, the company replaced phones with repaired devices instead of brand-new ones as it did in other countries.

The annual program featured a live studio audience, musical numbers, and elaborate sets. One song was set to the tune of Journey's "Don't Stop Believin'."

> *Life presents problems, please don't give up*
> *Let us maintain our rights*
> *Shed a smile and believe tomorrow will be better*
> *To repair life with a smile*

Known as the 315 Evening Gala, the more than two-decade-old program exposed business misconduct and celebrated consumer rights. Though local companies were also named, foreign firms were a popular target, particularly as the increasingly affluent market had risen in importance in the world. In the past, Hewlett-Packard, McDonald's, and French retailer Carrefour had fallen under scrutiny. Companies considered the impact of the show to be so profound that some of them went out of their way to resolve known problems in advance to avoid being named.

After the gala ran, Apple's Chinese fans mocked the report on the Internet, criticizing the network of using the attack to divert attention from more serious problems such as pollution and more egregious violations by domestic companies. Internet users also had harsh words for celebrities who had immediately chimed in against Apple on the Twitter-like Weibo service. A Taiwanese-American entertainer, Peter Ho, was heavily criticized when it appeared he sent out a message with a note to himself that he forgot to delete.

"#315isLive# Wow, Apple has so many tricks in its after-sales services. As an Apple fan, I'm hurt. You think this would be acceptable to Steve Jobs? Or to those young people who sold their kidneys [to buy iPads]? It's really true that big chains treat customers poorly. Post around 8:20."

Ho claimed that his account had been hacked, but the speculation was that he and other stars had been asked by CCTV to post the comments at a specific time.

Despite the outpouring of support, the controversy refused to die as other state-run media joined CCTV in their attacks. *People's Daily*, the official mouthpiece of the Communist Party, published a series of articles and editorials including one titled "Defeat Apple's Incomparable Arrogance."

The editorial charged that Apple had lost its integrity. Knowing that most Chinese admired Steve Jobs for his innovative spirit, the writer—a senior editor and party member—implied that Apple was neglecting the same people that held him in such high esteem.

The article also hit on deep-rooted resentments about Western companies and their exploitation of China.

"The profit-striving nature of capitalism has driven Apple to craziness," said the writer. "Chinese consumers have always felt powerless when confronted by the arrogance of these big Western brands."

More critics emerged. China's State Administration of Industry and Commerce called for "strengthened supervision" of Apple. Other television broadcasts followed, airing segments showing their journalists being turned away at Apple's offices. In one, a female reporter for financial news network CCTV Channel 2 was filmed in an encounter at Apple's Shanghai office.

Covering the camera lens with his hand, a man tells her she must stop recording.

"If you want an interview, you need to make an appointment."

"Who is available for an appointment?"

"It's all online. You can send an email."

"What if there's no reply?"

"That's his business if the email isn't answered."

When another staff member appeared, the female reporter tried again.

"Who shall we make an appointment with?"

"You need to find out the person on your own."

"I know that. What if my emails are not answered?"

"If we need an interview with you, we will contact you."

"So we have to wait until you initiate contact?"

A voice-over tells viewers that the network is still waiting for a response from Apple.

Some analysts and bloggers found the orchestration of the attacks to be bizarre. Regardless, the negative coverage posed a real problem for Apple.

Long-time China watchers believed they knew why Apple was targeted. The company had not spent enough time cultivating its relationship with the government and investing in efforts that benefited China and its people. Historically, Apple favored a one-size-fits-all approach in selling its products around the world and did little to tailor its business practices for individual countries. The company had moderated that stance in recent years, adding some customization in advertising and expanding its presence in key overseas markets. In China, Apple integrated Chinese Internet services into the iPhone and improved the way users could type in pinyin, the system used to spell Chinese words in Roman letters. But the efforts were minimal compared to other global corporations. On Apple's App Store, for example, Apple set prices across the world by predetermined tiers. Developers could choose one, and that determined the cost of their app in each country's currency. There was little flexibility for them to customize their pricing by market.

That was how the company had gotten into trouble in the first place. The one-year limited warranty that Apple provided in China was the standard policy in the United States; customers had to pay $99 for a two-year extended warranty. The company hadn't taken the country's regulation into account. In China, there was an expectation by the people that foreign companies would behave better than everyone else. To provide a service level that was seen as worse was a double betrayal.

Foreign businesses that had operated in China for a long time also implicitly understood that the Chinese government was happy to allow them in the country as long as these outsiders employed Chinese workers, paid Chinese taxes, and ideally shared knowledge and expertise. The one taboo was speaking publicly about the hoards of money the companies were raking in through these ventures.

Apple had crossed that line. Quarter after quarter, the company provided detailed information about their business in Greater China. That included Hong Kong and Taiwan, but the vast majority was in mainland China. Just a couple of months earlier, Cook had spoken of how in the past quarter the company had increased its revenues in the region by more than 60 percent to $7.3 billion as iPhone sales saw triple-digit growth. He also told analysts that Apple had eleven stores in Greater China, up from six the previous year.

"It's already our second-largest region," Cook said. "It's clear there is a lot of potential there."

These glowing reports played well in America, especially among Apple's stockholders. But in China, nobody wanted to hear about how much money a foreign company was making off the country. Every yuan that went into Apple's coffers was one yuan that wasn't going to a domestic company. China's flagship technology companies like Huawei, Lenovo, and ZTE harbored ambitions to someday become global leaders in mobile communications. It was in China's best interest to make sure that Apple didn't become too powerful in its home market.

"Apple is begging to be cut down to size as far as the government is concerned," observed David Wolf, a market strategist with more than two decades of experience in China. He added that Apple and China's interests had been aligned so far, but that was becoming less true. "Consumer day was the clearest signal that any of us could have been given that the balance is now tipped. . . . Had this just been CCTV and a few party interests going rogue, then it would not have intensified afterwards."

Apple did the only thing it could do to diffuse the crisis. On April

1, Cook apologized to the Chinese people in a lengthy open letter, written in Mandarin. In it he promised to be clearer about Apple's after-sale service and outlined an amended warranty policy, under which damaged or defective phones would be replaced with brand-new devices rather than refurbished models. Warranties would also be renewed for a full year from the date of replacement.

"We realize that a lack of communication in this process has led the outside to believe that Apple is arrogant and doesn't care or value consumers' feedback," he wrote. "We sincerely apologize for any concern or misunderstanding this has brought to the customers."

Cook closed the letter by thanking everyone for the valuable feedback. "We always bear immense respect for China and the Chinese consumers are always our priority among priorities."

Some bloggers interpreted the apology as a token gesture, but the appropriately contrite letter was otherwise embraced. Apple's image changed overnight at least for the moment.

"Its reaction is worthy of respect compared with other American companies," declared the *Global Times*, a tabloid published by *People's Daily*. The Foreign Ministry also praised Apple for "conscientiously" responding to consumers' demands.

Despite the humility so effusively expressed in the letter, Apple returned to its customary haughtiness when the U.S. media asked for a translation. Apple declined. The apology was only intended for Chinese consumers. Even by Apple's standards, the refusal was silly. Posted on Apple's site in China, the letter had already been made available to more than a billion people. Did the PR department really think they'd be able to control who else read it?

The pattern was the same, whether Apple was stiff-arming reporters, or squeezing its business partners until they bled, or neglecting to properly test its own products before they launched, or failing to appreciate the nuances of operating in other countries and other cultures.

Again and again, the damage was self-inflicted.

Nowhere was Apple's decline more obvious than on Wall Street.

Since Cook had taken over, Apple had continued to post strong earnings, but growth was slowing. After reaching a high of $702.41 the previous September, Apple's shares had plunged by nearly 30 percent to $500 by mid-January 2013. Shares fell even further after Apple announced the slowest quarterly profit growth since 2003. During the holiday quarter, Apple's profits had risen less than 1 percent to $13.1 billion. Sales rose by just 18 percent compared to the 73 percent growth recorded the previous year.

Particularly after the news of the slashed orders for iPhone 5 parts, Wall Street worried about the intensifying competition and a maturing market. Apple had sold iPhones and iPads to almost everyone who wanted one and was finding it more difficult to increase sales. The market was also moving from high-end models to the lower end, a category in which Apple did not have a product. In tablets, Amazon came after Apple with ads that compared its $299 Kindle Fire to the $499 iPad.

At the annual confab Cook addressed Apple's disappointing stock performance.

"I don't like it, either," he said. "And I can assure you the board doesn't like it and the management team doesn't like it. What we're focused on is the long term." Cook promised that the company was working as hard as ever on new products.

His comments did little to appease some investors. The first question came from a shareholder who wanted to know why Apple wasn't fighting Samsung and Android harder. "In the last few months, my shares have been down thirty percent in value," he said. "Why hasn't Apple used some of that huge cash to do total war for the market that we invented?"

Cook's response: Apple's goal wasn't to make the most devices, it was to make the best.

Apple's shares that day closed down 1 percent.

Cook's inability to instill confidence in the company was attracting notice. In a widely discussed article titled "The Last 6 Times Tim Cook Has Talked, Apple's Stock Has Dropped," *Huffington Post* correlated Apple's stock decline with Cook's public comments. The article was somewhat misleading because Apple's shares had been on a downward trajectory since September. The declines were also not always significant. But the article was yet another sign that Apple was losing its public relations campaign.

A couple of months later Apple reported its first profit decline in a decade. Particularly worrisome was the declining market share of the iPhone business. Many customers who had purchased an iPhone that quarter bought the older iPhone 4 models, which were considerably less profitable for Apple because they were cheaper.

Cook sought to placate investors by announcing plans to return another $100 billion to shareholders through higher dividends and stock buybacks, more than twice as much cash as he had announced the previous year. The news bumped Apple's stock up 5 percent briefly, but shares soon fell again. The stock was down more than 40 percent from its September high. A company that was once bullishly projected to hit $1,000 a share or more was trading at around $400.

The root all of the anxiety was that three years after the iPad's release, there was still no sign of the next world-changing magical product. Siri was a dud and Apple Maps a flop. The iPhone 5 was hardly the brand-new, exciting upgrade that it was promised to be. The MacBook Pros were slimmer and lighter and came with a better screen than ever before, but they were too expensive for most consumers. The latest iPod touches and nanos came in multiple colors with advanced features, but the business was increasingly insignificant to Apple's growth. iPods contributed no more than 5 percent of the company's overall revenues.

Despite Jobs's vows never to make a smaller iPad, Apple finally unveiled a 7.9-inch iPad mini in October 2012 along with an incrementally improved next-generation iPad. But instead of leading its rivals in disrupting the market, the company had waited until after many of its competitors had already captured much of the demand. By then Samsung had been selling the smaller format for two full years. When Samsung had first unveiled its seven-inch Galaxy Tab in 2010, Jobs had derided it as a "tweener"—too big to compete with a smartphone and too small to compete with an iPad. But Samsung, Amazon, and others had proved Jobs wrong. Consumers liked the portability. By introducing an iPad of a similar size, Apple was effectively conceding that Jobs had been wrong. The iPad mini drew attention to the fact that Apple was now chasing the competition rather than leading it.

When the iPad mini went on sale in early November, some of Apple's biggest stores in New York, Tokyo, and Seoul still commanded lines. But in Amsterdam, business returned to normal just two hours after opening. In Hong Kong, staff outnumbered the customers. Apple sold three million iPads and iPad minis in the first three days, but the excitement had waned considerably.

In numerous interviews and other public appearances, Cook promised great things to come.

"The boldness, the ambition, the belief that there are no limits. The desire among our people to not just make good products, but to make the very best products in the world—it's as strong as ever," he told the audience at the 2013 Goldman Sachs conference in February. "Our North Star is great product."

But his oft-repeated line was sounding tired. Rumors swirled about Apple's plans for a television, an Internet-connected watch, and a cheaper, smaller iPhone. But there were no announcements to prove his declaration. When Jobs had died, he had supposedly left behind a product road map for the next few years. But the iPhone 5 had been the last model in which he had provided detailed input, and the market was changing fast.

In the April earnings call, Cook told analysts that new hardware, software, and services would be forthcoming in the fall of 2013 and throughout 2014. But that was a long time away, especially when Samsung and other rivals were flooding the market with new models. A chorus of questions was being raised about Apple's ability to keep innovating.

Inside the company, morale had dipped. Employee compensation was tied enough to Apple's stock that moods tended to rise and fall with the fluctuations in share price. But the lack of new products was making the troops restless, too. Many people worked at Apple because they wanted to change the world. After inventing the iPhone and iPad, making the latest iteration of an already existing device just wasn't that exciting. Longtime employees began leaving or retiring. Under Jobs they had worked themselves to the bone, and over the years they had been rewarded with stock options. The current price levels were still higher than they had ever envisioned. They didn't need to work as hard anymore, especially if they couldn't see a further upside. Newer employees also had less incentive to stay because they had joined too late to benefit from the huge run-up in stock price over the last few years before the decline started. They had a better chance of seeking their fortune elsewhere, at a company that was still on an upward trajectory.

Even the employees who stayed were less hungry. Whereas Jobs used to recall his executives from vacations so frequently that they wondered if he did it on purpose, Cook respected his people's personal time. With more flexibility, people began taking vacations more freely. Executives bought vacation homes and expensive new cars. Eddy Cue—an avid Ferrari fan—joined the board of the Italian sports car maker. After Monterey Car Week—an internationally renowned car show for racing enthusiasts—the legendary Christian von Koenigsegg himself stopped by Apple's campus to show off a Koenigsegg Agera R, a $2.5 million beauty that could accelerate from a standstill to 186 miles per hour in 14.5 seconds.

Ive was seen more often in London, where he went to the Olympics, attended arts events, and was photographed at the Burberry Spring Summer 2013 Womenswear show. He also took on a side project, agreeing to design a single limited-edition camera for Leica for charity. In February he was spotted at the St. Regis hotel lounge showing off sketches of the camera design that he had in mind. The designer also appeared on the BBC's iconic *Blue Peter* children's program, where the host presented him with a gold Blue Peter Badge, and he in turn presented the host with a larger badge made of solid aluminum with the lab's CNC automated milling machine. His industrial designers were enjoying themselves more as well, investing in restaurants and taking exclusive trips. In early 2013, a few of them went to Baldface in British Columbia, where they went skiing and snowboarding via snowcats and helicopters.

What they gained in happiness, they lost in intensity. Just as Gautam Mukunda had predicted in his theory about business physics, forces were dragging Apple to the mean. The trappings of success were weighing the company down. Without Jobs's larger-than-life personality to referee the tensions, rivalries had intensified. Cook's reign set off a new round of jockeying. Some observers wondered if the CEO was purposely allowing conflict to see who bubbled up. The firings of Forstall and Browett only upped the stakes.

With Forstall gone, power was flowing toward Ive. In the *BloombergBusinessweek* interview, Cook couldn't say enough about his chief designer.

"I don't think there's anybody in the world that has a better taste than he does. So I think he's very special. He's an original," Cook said. "I love Jony."

Now that Ive had control over software designs, he had more responsibility than ever for the company's success or failure. But that also meant risking a breakdown in the equilibrium in product development that Jobs had so carefully calibrated.

"There is more to designing products than just what the product looks like. You have to think about the product holistically, about what's inside and what's outside," said Jon Rubinstein, Apple's former hardware executive. "Only when the right balance is achieved do you get truly spectacular products."

If Apple became even more driven by design than function, its products might look good but work poorly. Problems such as the iPhone 4 antenna fiasco were almost certain to repeat themselves. Industrial design was about surface beauty. Software design was about functionality. The disciplines actually required different ways of thinking. This was why it was rare for a designer to be good at both industrial design and user interface. Ive could be a brilliant exception that understood that balance, but some people who had worked with him weren't so sure.

When Apple was working on the remote control for the Apple TV digital media receiver several years before, Ive and his designers had insisted that the controller have just six buttons: up, down, left, right, enter, and menu. Engineers argued that the remote should have at least a play button and universal volume buttons, so users could adjust the volume on their televisions while watching a show. Without those features, they would inevitably switch to another remote control. But industrial design won the argument. Though Apple later added a play/pause button, the first generation of Apple TV remotes shipped without either function. Because of the insistence on a minimalist design, users' interaction with Apple TV was made more difficult, not less.

The only way to prove the skeptics wrong was to come out with a ground-changing product. As Ive got serious about proving himself, he read his designers the riot act. No more playing. It was time to get back to work.

On a cold afternoon in March, reporters lined up for hours in New York City to attend Samsung's launch event for its latest Galaxy device. Five hundred thousand people tuned in to YouTube to watch remotely. Samsung was mounting a glitzy show, produced by Broadway veterans, at Radio City Music Hall.

The invitation to the launch did not reveal what kind of device Samsung was launching. "Come and meet the next Galaxy," was all Samsung had said in white print on a black card. Like Apple, the company didn't name the next device but teased the audience with the words, "Ready 4 the Show." Since the previous Galaxy was an SIII, this one was expected to be an S4. The blogosphere eagerly speculated on whether the next Galaxy might include eye tracking, a first-of-a-kind technology that would allow users to scroll through pages with eye movements.

"It's usually Apple, not Samsung, that gets this kind of attention," the All Things Digital blog noted. All of the attention had gotten rivals so nervous that LG took out a billboard on Times Square that played off the "4" in Samsung's ad right underneath it.

"It'll take more than 4 to equal one LG Optimus G."

HTC, meanwhile, crashed the party. Standing outside of the event hall, company staff demoed its new HTC One phone to those who had arrived early.

Apple wasn't above such ploys. Usually the company avoided media attention, turning away press requests with a blanket insistence that it never granted interviews. But on the eve of Samsung's launch, the public relations department was contacting reporters and offering up phone interviews with marketing chief Phil Schiller even though he had no new product to talk about. In the past Apple had provided a sneak preview of its next operating system in mid-March, but this year the company wasn't ready to even do that.

"We're in uncharted territory," a spokeswoman admitted.

Hustling one of its top executives was a feeble attempt to steal Samsung's thunder. Even though Schiller had nothing in particular to talk about, he did what he could. In an interview with the *Wall Street Journal*, he played down the competition and listed all the ways Android-based phones like the Galaxy were inferior to the iPhone.

"Android is often given as a free replacement for a feature phone, and the experience isn't as good as an iPhone," he told two *Journal* reporters. "When you take an Android device out of the box, you have to sign up to nine accounts with different vendors to get the experience iOS comes with. They don't work seamlessly together."

Schiller also presented Apple's own research showing that four times as many iPhone users switched from an Android phone than the other way around.

"The most important thing to Apple," he said, "is that people love our products."

The reporters wanted to return to Schiller's claim about how much consumers preferred iPhones to Android phones. If that was true, one asked, why was Samsung gaining market share?

"I don't think market share is the best measure," Schiller said. Though he was clearly trying to avoid mentioning Samsung, he seemed unable to help himself. Off the record, he suggested that Samsung and Android were picking up sales from the low-end market. "It's not coming at our expense as much as it was at other people's expense."

The reporters pressed. Why, they asked, was growth flat?

Schiller didn't want to answer.

"That's a question for our earnings call."

When was the next iPhone coming out?

Again, Schiller deflected, but this time with a whopper.

"Most customers are interested in the newest products and the products we've announced over the last few months," he insisted. "I think most people in the world aren't worried about what's coming."

The desperation of that statement was so naked that it took the breath away. Here was a company that had staked its fortunes upon its matchless ability to make the world stand in line for whatever jaw-dropping machine it invented next. Now that Apple's wellspring of innovation appeared dry, its marketing chief was reduced to claiming that the world didn't care.

Holy Grail

In the end, Apple needn't have worried about Samsung hogging all the attention.

Despite the buzz, the Galaxy S4 launch proved to be a colossal bust. The one-hour extravaganza, complete with a live orchestra, was a disaster. The show opened with a video of a little boy dressed in a bow tie as he tap-danced from his home into a Rolls-Royce to deliver the new phone up to the stage. From start to finish, the production felt strangely out of touch. Samsung was trying both too hard and not hard enough. The master of ceremonies, Broadway star Will Chase, looked like he wanted to flee the stage as his jokes and banter fell flat before the dumbstruck audience. J. K. Shin, the executive who headed Samsung's mobile communication division, strolled into the spotlight, triumphantly holding out his arms and inviting adoration as though he thought he was Elvis or Steve Jobs. But when Shin opened his mouth to brag about the new phone, he sounded stiff and awkward.

What had led Samsung to believe that stilted Broadway skits, featuring hackneyed characters and tone-deaf dialogue, would help sell their new product? Though a Tony-winning director was listed in the credits, Samsung executives in Seoul had micromanaged even the tiniest details of the production, down to the socks worn by the actors. Their misreading of modern American culture was staggering, especially in a skit near the end where several women at a bachelorette party—all of them holding a Galaxy—were shown worrying about their nail polish drying, joking about marrying a doctor, and ogling a shirtless gardener.

"Okay," said Chase, ushering them offstage. "I think you girls are done."

The launch hadn't even ended before Samsung fell under attack. Many decried the company's retrograde take on women.

"Samsung weird," declared a headline on the *Verge* site. "How a phone launch went from Broadway glitz to sexist mess."

"I don't get offended very often," wrote tech blogger Molly Wood. "But Samsung's long parade of '50s-era female stereotypes, in the midst of an entirely long parade of bad stereotypes, just put me over the edge. Oh, they announced a phone? You'd barely know it."

For all the headway that Samsung had made in its TV commercials, the launch showed how the company still had a long way to go before it could hope to usurp Apple's iconic status. But as embarrassing as the event was, in the end it made no difference. The Galaxy S4 sold almost twice as fast as the previous model. Sales in the first month hit ten million, putting Apple on the defensive.

With no new product to compete with Samsung's latest device, Apple responded in the only way it could, claiming superiority through a marketing slogan.

"There's iPhone. And then there's everything else."

Apple's woes were deepening. Despite Tim Cook's prime seat at the State of the Union address, doubts were being raised about the company's patriotism. The year before, when the *New York Times* published its iEconomy series about Apple's effect on the U.S. economy, the newspaper had blamed Apple for moving manufacturing jobs out of the country and squeezing the middle class. One of the quotes in the article had garnered much attention because it was so smug.

"We don't have an obligation to solve America's problems," an unnamed executive had told the reporters. "Our only obligation is making the best product possible."

The article created such a furor that the company felt compelled to commission a study quantifying the number of American jobs that the company had helped create as a result of its success. According to its findings, Apple had created or supported more than five hundred thousand jobs, more than ten times the number of people it directly employed.

The *New York Times*'s dissection of Apple had been unrelenting. A few months later the newspaper published another powerful exposé focusing on how the company was sidestepping taxes by setting up shell offices in Nevada and overseas, where tax rates were much lower than in California. Describing an accounting technique called "Double Irish with a Dutch Sandwich," the newspaper detailed how Apple routed profits through Irish subsidiaries and the Netherlands and then to the Caribbean. Without such tactics, Apple would have had to pay $2.4 billion more than the $3.3 billion it paid in 2011. At a time when the government's cash was running short and federal programs were being cut, the notion that large corporations were avoiding taxes was unforgivable.

By the time the articles won a Pulitzer Prize in April 2013, the idea that Apple was skirting billions of dollars in taxes—and was contributing to the nation's economic downturn in other ways—had been firmly planted in the national conversation. In an interview with *BloombergBusinessweek*, Cook was asked about Apple's obligation to the country.

"I do feel we have a responsibility to create jobs," said the CEO. "I think we have a responsibility to give back to the communities, to pick ways that we can do that . . . and not just in the U.S., but abroad as well. I think we have the responsibility to make great products that we can recycle and that are environmentally friendly. I think we have a responsibility to make products that have a greater good in them."

Inspiring as that answer sounded, Cook's assertion of Apple's higher purpose was not so easily squared with the revelations of the company's tax avoidance schemes. How exactly did the Double Irish with a Dutch Sandwich serve the greater good?

Cook was about to be asked the question again, this time in front of Congress. For years, the Senate Permanent Subcommittee on Investigations had been trying to nail technology companies that were taking advantage of the country's outdated tax code to pay little to no taxes. After skewering Microsoft and Hewlett-Packard over their accounting tactics the previous fall, the committee was now turning its attention to Apple.

At issue were the lengths to which Apple went to squirrel away its profits overseas. Of about $145 billion in cash, cash equivalents, and marketable securities, $102 billion was being kept outside the United States, untaxed by the company's home country.

Until then, Apple had kept a low profile in Washington in part because Steve Jobs had never had patience for dealing with the government. Among the tech giants, its spending on lobbying efforts was one of the lowest for the company's size—$9.05 million since 2008, compared with $38 million for Microsoft and $38.2 million for Google, according to the Center for Responsive Politics. But as Apple's profits grew, dealing with the government had become unavoidable. Its movements were followed and appraised more closely.

The Senate subcommittee's questions could not be ignored. If Apple wanted to remain successful, its leadership needed to maintain at least a façade of goodwill toward the government. That spring as hearings approached, the company cooperated with the subcommittee investigators and made Apple executives, including Cook, available for interviews. When the panel asked the CEO to testify at a hearing as well, he agreed. Better to show up willingly than to be subpoenaed.

Apple also hired a lobbyist who had founded Microsoft's Washington office and retained O'Melveny & Myers, a law firm experienced in shepherding large companies through the minefields of regulatory scrutiny. The firm's clients included Enron and Goldman Sachs. Among its partners was George Riley, the longtime friend of Jobs who had counseled Apple before on legal matters.

In high-profile congressional hearings such as this, where a corporation's national reputation often hung in the balance, a company could gain strategic advantages by working closely with the committee in advance of the public grilling. Typically, the company's lawyers or other representatives sought to limit the scope of the investigation and also to assess the lines of questioning their clients might face. Knowing in advance where the most devastating punches might land allowed the company to prep for the big moment. The committee had invited three of Apple's executives to answer the senators' questions—Cook, as well as the company's chief financial officer and the head of its tax operations. In the weeks leading up to the hearing, the trio was put through intense coaching sessions by O'Melveny & Myers's crack team. Every conceivable question and scenario was thrown at them until their answers struck the desired note—firm, yet deferential. Cook and the other witnesses needed to prepare just as they would for a court proceeding. Everything they said would be recorded. Every word could be used against them and Apple later.

A week before the hearing, Cook set the stage as he gave interviews to *Politico* and the *Washington Post*, two of the most widely read publications inside the Beltway. The committee was expected to release the findings of its investigation any day, and Apple wanted to get its side of the story out first. Instead of merely defending his company against charges of tax dodging, Cook used the interviews to pitch a simplified corporate tax law.

Everyone understood that the committee's work was just another type of political theater. For all the senators' blustering, the Democrats and Republicans were unlikely to agree on any legislation to close tax loopholes, especially in the midst of the capital's toxic gridlock. Nor were they likely to follow whatever suggestions Tim Cook offered on streamlining the tax code. But by introducing his proposals ahead of the hearing, Cook increased the odds that he would appear proactive rather than defensive.

"If you look at it today, to repatriate cash to the U.S., you need to pay 35 percent of that cash. And that is a very high number," Cook told the *Post*. "We are not proposing that it be zero. I know many of our peers believe that. But I don't view that. But I think it has to be reasonable."

He also planted the idea that Apple was already doing plenty for the country.

"Apple is paying approximately one million dollars an hour in just domestic income taxes," he said. "You may not know this, but Apple likely is the largest corporate taxpayer in the U.S."

The day before the hearing, the committee presented the results of its months-long investigation. The report charged that Apple had structured organizations and business operations to avoid paying U.S. taxes on overseas income. It alleged that Apple transferred vast amounts of its assets and profits to subsidiaries in Ireland that didn't have a tax residence status anywhere in the world. One holding company, Apple Operations International, made up thirty percent of Apple's total worldwide net income from 2009 to 2011 but did not pay any corporate income tax to any national government during that period. Another company, Apple Sales International, did file a corporate tax return related to its operating presence in Ireland, but the subsidiary paid little to no taxes on that amount per its agreement with Ireland. The panel suspected that both shell corporations were set up to exploit the difference between Irish and U.S. tax residency rules. While Apple claimed effective tax rates between 24 and 32 percent, that percentage included U.S. state and foreign taxes. When those were separated out, Apple's U.S. corporate income tax was 20.1 percent, compared to the federal statutory rate of 35 percent.

"Apple sought the Holy Grail of tax avoidance," said the committee chairman, Senator Carl Levin, in a statement released with the findings. "It has created offshore entities holding tens of billions of dollars, while claiming to be tax resident nowhere. We intend to highlight that gimmick and other Apple offshore tax avoidance tactics."

Senator John McCain, the ranking committee member, also accused Apple of being one of America's largest tax avoiders, calling Apple's tax structure "byzantine." The two congressmen were on opposite sides of the aisle in their politics, but they were joined in their indignation. Every expectation was that the hearing would be confrontational.

On the morning of May 21, 2013, Cook arrived at the Dirksen Senate Office Building in a perfectly tailored dark suit with a white shirt and solid blue tie.

From the beginning, the senators made it clear that Apple had not broken any laws and that they were studying the company's tactics to figure out how they could close tax loopholes and update the tax code. Even when they criticized Apple, their comments were couched with respect. Several expressed their love for the company's products. McCain made a point of noting that Cook was "an outstanding executive" even as he accused the company of indirectly increasing the general public's tax burden.

As with any hearing on Capitol Hill, the discussions were just as much about the politics among the senators as it was about the issue. The libertarian senator, Rand Paul, quickly rose to Apple's defense.

"I'm offended by a four-trillion-dollar government bullying, berating, and badgering one of America's greatest success stories," Paul said. "What we really need to do is apologize to Apple, compliment them for the job creation they're doing, and get about doing our job. Look in the mirror and let's make the tax code better, fairer, and more competitive worldwide. Money goes where it's welcome. Currently our tax code makes money not welcome in this country."

Levin responded icily.

"You're of course free to apologize if you wish."

Apple had always considered itself to be a progressive company, and Cook's personal politics were more aligned with Levin's than Paul's. But if the CEO saw the irony of the situation, he didn't show it.

Apple had released the prepared remarks the previous day in response to the committee's report, but Cook read it again so his words could be recorded in the Congressional Record.

The CEO sounded a patriotic note, emphasizing the contributions that the company had made to the U.S. economy. He mentioned how Apple's U.S. workforce grew fivefold to fifty thousand in the last decade and reminded the senators of the company's plans for a Mac product line in the country. He also explained that the overseas cash funded the company's foreign operations.

"We pay all the taxes we owe—every single dollar," Cook said before throwing Levin's words right back. "We don't depend on tax gimmicks."

He quoted his favorite line from President Kennedy, "To whom much is given, much is required," to stress his certainty that he and Apple were doing their part for the country. He also formally presented his tax proposal: a revenue-neutral code that lowered corporate income tax rates and set a reasonable tax on foreign earnings. Cook even managed to connect the simplicity of his plan with Apple's philosophy in developing products: We believe in the simple, not the complex.

"We make this recommendation with our eyes wide open, realizing this would likely increase Apple's U.S. taxes. But we strongly believe such comprehensive reform would be fair to all taxpayers, keep America globally competitive, and would promote U.S. economic growth."

So far Cook was performing his role superbly. The main spectacle was the question-and-answer session in which the senators were given the chance to grill Apple's executives. But for all of the drumbeating they had done before the hearing, here too the senators seemed to waver between adoration and condemnation. Whatever behind-the-scenes discussions had taken place between the two sides in advance of this day, Apple had been successful in shaping the debate. There was little mention of how the company had overstated its effective U.S. income tax rate by including state and foreign taxes into the calculation.

Before McCain began his questioning, he congratulated Cook for his successes. "I think you'd have to be a pretty smart guy to do what you do and a pretty tough guy, too," he said before asking Cook if he felt bullied or harassed by the committee. McCain's tone was mild, even friendly.

"I feel very good to be participating in this and I hope to help the process," Cook replied earnestly as he leaned forward. "I'd really like for comprehensive tax reform to be passed this year and any way that Apple can help do that, we are ready to help."

Cook told the committee that he had come willingly.

"I think it's important that we tell our story and I'd like people to hear it directly from me."

"So you were not dragged before this committee?" asked McCain.

"I didn't get dragged here, sir," Cook said with a grin.

"You don't drag very easily, I understand," the senator said as Cook laughed heartily.

When McCain became serious and asked if Apple had an unfair tax advantage, Cook gently disagreed.

"No, sir, it's not the way that I see it and I'd like to describe that."

The CEO explained how Apple was paying all the taxes it owed for income generated in the United States.

In the end, McCain drew back his fire.

"What I really wanted to ask was why the hell I have to keep updating the apps on my iPhone all the time and why you don't fix that?"

"Sir, we're trying to make them better all the time," Cook told him, grinning again.

Throughout the questioning, Cook took care to address each senator respectfully, addressing them as "Sir" or "Senator," in the case of the women. He thanked them for asking questions and stressed their shared values and mutual love for their country. He also validated the senators' questions and opinions at every possible chance.

"How important do you think it is that we change the tax code to ensure that this remains a good place for investment?" asked Senator Kelly Ayotte.

"I think it's vital to do," replied Cook. "I think it's great for America to do. I think we would have a much stronger economy if we did that. I think it would create jobs and increase investment. And so I put my whole weight and force behind it."

When Ayotte suggested that more jobs and investment could also result in increased tax collections, Cook readily agreed.

"I think that's a very excellent point . . . all ships rise with the tide."

While the coaching he had received was evident, the senators also saw a glimpse of the sharp, analytical mind that drove Cook's subordinates to achieving excellence. Though it was unlikely that the longtime Apple executive had been directly involved in the nitty-gritty of Apple's tax strategy, he had absorbed every last detail by this time. The senators, who had a much more general grasp of the picture, were no match.

Cook was also careful to maintain an open, friendly, and noncontroversial demeanor. Though all three Apple executives went out of their way to exhibit deference, any responses that could be remotely construed as combative or challenging were given by his chief financial officer, Peter Oppenheimer, and tax strategy chief, Phillip Bullock.

When Senator Rob Portman mentioned how Samsung's global tax rate was about the same as Apple's, it was Oppenheimer who stepped in to point out that Apple was at a bigger disadvantage because it was not able to freely move its capital back to its home country like Samsung could because of the high U.S. tax burden.

Only at the very end did questioning become tense. Levin asked Cook if he was correct in his understanding that Apple had signed an agreement in 2008 that effectively shifted much of the economic rights to its intellectual property—its crown jewels—to Ireland by keeping the overseas profits there.

"I think we ought to get a straight answer on this," said Levin.

Cook interrupted him. "I would disagree with your characterization."

"Well, you signed the agreement in 2008, didn't you? Don't you work for Apple?"

Oppenheimer tried to step in for his boss, but Levin refused to let go.

"Mr. Cook, you signed that agreement, did you not, in 2008?"

"I signed the 2008 agreement, yes."

"You were working for Apple at that time?"

"I've been working for Apple for fifteen years, sir."

"Three people working for Apple signed this agreement," Levin concluded. "Mr. Cook, I couldn't disagree with you more. Of course you shifted something—the most valuable thing you have—the economic rights in your intellectual property."

A few minutes later, Levin asked Cook, "Is it true you told our staff you're not bringing the hundred billion home unless we reduce our tax rates? Is that accurate?"

"I don't remember saying that," Cook said, evading a straight answer for the first time.

Levin pressed him again. "Is it true?"

Cook's voice grew testy. "I said I don't remember saying that."

"No, I'm saying is it true that you're not going to bring them home unless we reduce our tax rates?"

After the third time, Cook paused and looked up from the corner of his eyes as if he were rapidly sifting through his mind for the right answer. "I have no current plan to bring them back at the current tax rate."

"All right," said a frustrated Levin. "Is that the same way as saying unless we reduce our tax rates, you're not bringing them home? Is that the same way?"

"No, I don't think it's the same, sir."

"How is it different?"

"Your comment sounds like it's forever and I'm not projecting what I'm going to do forever because I have no idea how the world may change."

The session ended with the senator launching into a monologue about the need to change the tax code.

"We're proud of you being an American company. We're glad you're where you're at, but the result of these arrangements that you've continued is that most of your profit is now where we've described all morning, in Ireland, in these companies that don't exist anywhere except on the water. . . . Of course we've got to change this system, but in order to change it, we've got to understand it, not deny it."

Still, when the reviews of Cook's performance came in, they were almost unanimous. He had outdone himself. He had remained calm and controlled. He had shown himself to be authoritative yet respectful. Most importantly, he had held his ground even when Levin tried to bait him. By doing so, he had dominated the hearing and averted a headline-making showdown. Cook had even impressed the senators. In a complete about-face from his position before the hearing, McCain's spokesman told reporters that the senator didn't want regulators to punish Apple.

At the witness table on Capitol Hill, Cook had finally shown himself to be the CEO that he could be. He might not have Jobs's dynamism, but he exhibited a different kind of gravitas and wit. Sitting across from the senators in his tailored suit, he looked like he belonged. He was one of them. He had done what's right.

As positive as the reactions were in the immediate aftermath, the hearing took some more shine off of Apple's mystique. The company may have had the legal right to set up the tax structure the way that it had, but in the end, Apple had made a self-serving decision to keep most of its cash overseas to avoid paying taxes. It was no more than what many other companies did, but Apple stood out because of its immense success and its claim that it existed to make the world a better place.

On the same day, George Packer, a journalist for the *New Yorker*, published a book in which he discussed the changing ethos in America, where individuals are out for themselves and "winners win bigger than ever, floating away like bloated dirigibles, and losers have a long way to fall before they hit bottom, and sometimes they never do."

In interviews, Packer singled out Apple as one of the worst offenders because the company pretended to be otherwise. "Apple styles itself as a lifestyle, almost as a revolution. This slogan from the nineties Think Different suggested that Apple is like Gandhi or like King. It's a liberating movement," he told San Francisco's public radio station KQED. "Apple is a company . . . they are just another company."

Within a few days, Apple was in the hot seat yet again. The Justice Department's case accusing Apple of masterminding a monopoly to increase digital book prices was set to go to trial on June 3.

According to the government, the case was straightforward price-fixing. Before Apple opened its digital bookstore, publishers had sold books to retailers at a wholesale price. But when Amazon sold many of the books at a loss to shore up their dominant position, the publishers had schemed together to stop the price erosion. Apple was accused of serving as the "ringmaster" by brokering a deal that allowed the publishers to dictate the price in its digital bookstore in exchange for a 30 percent cut of the revenues. The assurance of having a major retail partner then allowed the publishers to impose the same agency agreement on their other retailers. Apple also inserted a most-favored-nation clause requiring that prices in other stores, like Amazon's, be matched in Apple's bookstore. As a result, book prices rose. The government had plenty of evidence showing the communications that took place among the publishers as well as between them and Apple. All five publishers that had been named in the suit had already settled and had agreed to terminate the existing agency agreements.

Apple contended that it had not known the extent of the publishers' collusion with each other. They had been discussing strategies to deal with Amazon's cut-rate pricing even before Apple had decided to start the iBookstore. The company also asserted that Apple couldn't be accused of violating antitrust laws because it was a latecomer to the industry, not a dominant player. It was simply trying to come up with a viable business model that would introduce competition in the market.

The fate of the case had the potential to impact not only Apple and Amazon, but also consumers, booksellers, authors, and literary agents. When the first three publishers had settled, the court received 868 comments on the proposed final judgment from individuals, companies, and industry groups, including a consumer activist group. Among them was an amicus brief from a lawyer in the form of a five-page comic strip that argued against the settlement. More than 90 percent of the comments had been negative, critical of Amazon's cut-rate pricing and full of praise for the entry of new competitors as a result of the agency agreements.

Ten days before the trial was set to open in New York, attorneys from both sides met with U.S. District Judge Denise Cote, a veteran judge and former prosecutor with a flair for literary references.

Cote was familiar with the case because she had approved the settlement agreements. In September 2012, at the time of the settlement with Hachette, Simon & Schuster, and HarperCollins, she had quoted a poem in her opinion to stress the importance of books.

"There can be no denying the importance of books and authors in the quest for human knowledge and creative expression, and in supporting a free and prosperous society. To quote Emily Dickinson: 'There is no Frigate like a Book / To take us Lands away, / Nor any Coursers like a Page / Of prancing Poetry— / This Traverse may the poorest take / Without oppress of Toll— / How frugal is the Chariot / That bears a Human soul.'"

Toward the end of the hearing, a lawyer for the Justice Department asked if Cote might share her thoughts on the case based on the evidence. Unlike the patent trial against Samsung, this would not be a jury trial but a bench trial decided by the judge herself.

Surprisingly, the judge acquiesced. While she stressed that her view was still tentative and based only on the emails and correspondences among the companies named in the conspiracy, she indicated that she was siding with the Justice Department.

"I believe that the government will be able to show at trial direct evidence that Apple knowingly participated in and facilitated a conspiracy to raise prices of e-books," she said, "and that the circumstantial evidence in this case, including the terms of the agreements, will confirm that."

The judge promised to keep an open mind, but the statement was worrisome for Apple. Before the trial had even begun, Cote had already begun writing a draft of her decision.

19

The Red Chair

In late May 2013, on the opening night of the *Wall Street Journal*'s annual three-day tech gathering, Tim Cook walked onto a stage in front of hundreds of his peers and settled into a bright red leather chair.

The All Things Digital conference was an invitation-only event hosted by Walt Mossberg and Kara Swisher, two of the biggest names in technology reporting. The speakers were mostly A-list executives from the hottest companies. With a no-nonsense style, Mossberg and Swisher questioned executives in front of more than six hundred of the smartest and most influential players in the industry. No speeches or slide presentations were allowed.

For years, Steve Jobs had sat in the same red Steelcase chair, drawing the audience into his reality distortion field with his always entertaining and sometimes misleading insights about Apple, its competitors, and the industry. In 2007, he flashed the iPhone that Apple was about to put on the market. "Best iPod we've ever made. . . . Best phone we've ever made, too," he quipped as the audience laughed. At his last conference in 2010, Jobs spoke about everything from Apple's products to controversies such as the missing iPhone 4 prototype and the lack of Flash support. He held the audience spellbound, tantalizing them with nuggets such as how the iPad development started before the iPhone and how Apple was structured like a start-up.

Cook's first appearance was in 2012, nine months after he had been appointed CEO. Both the hosts and the audience were warm and supportive as he sounded an optimistic note in his first extensive interview. "It's an absolutely incredible time to be with Apple," he said, "and I'm loving every minute of it." Cook had been thoughtful and genuine in his enthusiasm. Attendees were also pleasantly surprised by his sense of humor.

This year, as Cook sat in the red chair again, the tone of the conversation was more pointed. Before the CEO could catch his breath, Mossberg was cataloging the problems now clouding Apple's horizon, starting with the challenges from Samsung and Android, then moving on to the controversies in Congress and in China.

"You're being beaten up by various governments over various things, and your stock is down—down significantly. . . . I actually don't know the price of the stock, but whatever. There's a sense that you may have lost your cool, that someone else has got the cool—that Samsung has got the cool."

Mossberg then asked the question on everyone's mind.

"Is Apple in trouble?"

"Absolutely not."

Cook was smiling, but his delivery was unconvincing. He fell back into the company line about how iPhone and iPad owners used their phones and tablets more than other device owners. He spoke of the "unprecedented" number of new products Apple had released the previous year, but neither his voice nor his face showed any emotion.

"I feel pretty good. I feel pretty good," he claimed, but his voice dropped weakly.

Swisher refused to let him off the hook. In the tech world, she said, there was a growing sense that Apple was on the decline. "You have a lot of really tough rivals compared to before."

"We've always had competent rivals," Cook countered, citing Microsoft and Dell. He put his hands thoughtfully together as he returned to another oft-repeated line. "Our North Star is always on making the best products."

"So the outside perception doesn't bo—?" asked Swisher. "Because it's there. It's clearly there and you hear it more and more. People are worried for Apple."

Cook said the perception did not trouble him. He acknowledged that the declining stock price was frustrating, but he reminded Swisher that it wasn't unprecedented. "I guess the beauty of being around for a while in this is that you see many cycles," he said, trying to sound philosophical. "If we make great products that enrich people's lives, then the other things will happen."

Mossberg and Swisher turned their attention to products. As Cook started to defend the iPad mini, saying that innovation didn't have to mean creating a whole new category of devices, Mossberg cut him off. "I'm not trying to do a wordplay here," he said. When, he asked, was Apple coming out with the next big thing? Where was the television? What about wearable devices like Internet-connected glasses and watches?

"I don't want to go into detail," said Cook, deflecting. "I don't want to answer those. . . . I have nothing to announce." Jobs had never talked about upcoming products, either, but at least he had always had something interesting to say about the company's creative process. He knew how to intrigue an audience without actually revealing anything new.

Didn't Cook worry about competition from Android?

"I don't have my head stuck in the sand," he said, referring a second time to how much time people spent using Apple's products. Every time the interviewers pressed him on Apple's plummeting market share, Cook tried to change the subject. As the minutes ticked by, the smile on his face grew tight.

The hard questions just kept coming. Was the amount that people actually used Apple products the only meaningful metric? Did sales figures of Apple devices no longer matter? What about customer satisfaction? Young people, Swisher pointed out, weren't using iPhones as much anymore because they were bored by the look and feel of the device. Was Cook worried about the graying of Apple's demographic? Was the company working on a redesign to appeal to younger consumers?

"Our customers are all ages," said Cook, "and I love that."

By now he was running out of ways to say nothing. His answers began to strike the same note. Apple was doing great, he insisted. Exciting new products were coming down the pipeline, he promised. His confidence in the company's future, he said, had not wavered. Everyone on his team was doing fabulous work, he said. The culture inside One Infinite Loop was unchanged. Everything was still hunky dory.

Cook was looking away now, barely glancing toward the audience. He was fidgeting with his watch. Had he ever been backed into a similar corner, Jobs might well have exploded. Cook remained unfailingly polite, even as his entire body went rigid. Only a few weeks before, in preparation for his testimony before Congress, he had been coached on how to handle an interrogation. Working hard to sound logical, he slowed down his answers and quoted numbers and statistics. He stuck to the script, fending off any attempt to lead him off message. Such tactics had worked well when faced with aging senators whose understanding of technology was minimal. But this audience was not so easily fooled. This was a savvy Silicon Valley crowd. In this room, Cook's answers fell flat.

Asked about the upcoming e-book trial, the CEO tried to trivialize the government's case. He said that Apple had done nothing wrong, that it had conducted its dealings with the publishers with complete propriety, and that in the end he and his company would be vindicated. On a question about why Apple wasn't making better use of its piles of cash by buying companies, he argued the premise of the question rather than responding directly. Just about the only admission he was willing to make was that the company had screwed up with Apple Maps, which was not only self-evident but also a statement he had made before.

The performance was a disaster. Cook came across as delusional and painfully out of touch. If he was truly unfazed by the host of problems facing Apple, if he actually believed that everything was going well, then the company was really in trouble.

A particularly worrisome moment came near the end, when Dan Benton, a well-known tech-focused hedge fund manager, stood up and made a pointed observation. Cook's comments, Benton noted, sounded like the kind of rationalizations that Apple had resorted to in the nineties, before Jobs had returned, when the company was on the verge of going under.

"Why won't you let us dream? The guys at Google are presenting a world right now of gigabit fiber and weather balloons that do wireless and Google Glass," he said. "Why won't you give us a glimpse of the future as Apple sees it?"

Benton was practically begging the CEO to share some hint of a vision. Cook's reply was so dispassionate that it sucked the air out of the room.

"From at least the last fifteen years, we've done the same thing," he said, speaking almost by rote. "We release products when they are ready, and we believe very much in the element of surprise, and we think customers love surprises, so I have no plans on changing that."

Even before Cook left the stage, audience members were sending out withering critiques.

"If 'incredible' was tweeted trigger for Tim Cook drinking game, #d11 would be an alcoholic toxic disaster," tweeted *Wired*'s senior writer Steven Levy. Tonya Hall, whose news program aired on Bloomberg Radio, tweeted: "The leadership gap left by Steve Jobs looks widest when Tim Cook is on stage talking about the company."

Some of the sharpest criticism came from *Fortune* senior editor Adam Lashinsky, whose coverage of Cook's regime had been mostly favorable until this moment.

"It is a strange sight to see the CEO of Apple, a company known for its brilliance and vision, decline over and over to discuss just about anything in any detail," he wrote, saying that reporters were grasping for kernels of news. The journalist called the interview "alternately painful and tedious."

One of the toughest questions at the All Things Digital conference had been about Apple's ongoing patent war.

"You've been involved in patent litigation on Android for several years now and as far as I can tell, it hasn't really accomplished you anything, right?" said Nilay Patel, managing editor of technology site *The Verge*. "When does this end for you? What is the endgame?"

Cook had no good answer.

"I'm not negotiating it this evening," he said. "It's a values thing. This is about values at the end of the day."

More than three years after Apple began its war against Android, the company seemed no closer to a resolution. Its lawsuit against HTC had been settled, but it was still waging a two-front war with Motorola and Samsung.

In Apple's dispute with Motorola, the preeminent Judge Richard Posner had thrown out what had been expected to be a big court case in Chicago because it didn't make economic sense. He argued that advances in the software industry didn't require the kind of big up-front investments that justified legal protections and that companies benefited enough from being first to market with a technology. Appeals on that ruling as well as other cases around the world were winding their way through court.

Meanwhile, the fight against Samsung raged on. Negotiations had continued through the spring. A memorandum of understanding for a potential settlement had been drawn by representatives for Apple and Samsung and sent to senior managers on both sides. But so far none of those efforts had amounted to anything.

Publicly, Apple was taking a bullish stance. Its claims against Motorola and Samsung were received with mixed results, but its defense strategy was more successful. Apple had launched an aggressive public relations campaign against the standard essential category of patents that it was accused of infringing. The company also appealed to various government agencies and telecommunications bodies to set basic licensing standards, saying that companies were overvaluing patents that they were obligated to license on fair and reasonable terms. As a result, Apple had successfully defeated the majority of standard essential patent claims by Samsung. European regulators were also investigating.

But as meaningful as these rulings were, they didn't protect the key innovations that were supposed to differentiate iPhones and iPads. Every day that the court cases dragged on was a victory for Samsung and Motorola, who could keep selling their devices. That potentially had broad implications for Apple's future business. Android's collective market share gains would allow Google, not Apple, to control the burgeoning digital content business.

Throughout the spring and summer, Apple won some battles and lost others. In early June, the U.S. International Trade Commission sided with Samsung on a standard essential patent that covered technology used to send information over wireless networks. The ruling barred the importation of certain iPhone and iPad models that were designed to work on AT&T's networks. The trade commission rejected Apple's argument that Samsung couldn't use the patent to seek an injunction and said it believed Samsung's proposed royalties for licensing its patents were reasonable and offered in good faith. The commission also criticized Apple for using Samsung's patents without compensating the company under the guise that the licensing terms weren't fair or reasonable.

The impact of the ban was not huge. The devices in question were all older models, and the only one with meaningful sales was the iPhone 4 version for AT&T. Piper Jaffray estimated that the ban could cost Apple about $680 million, or 1 percent of its revenues for the next two quarters. But Apple appealed to the Obama administration to disapprove the ban before the order took effect. If it was granted, it would be the first time in almost thirty years that a president had intervened in an import ban.

On the morning of June 3, 2013, the Manhattan courtroom of the U.S. District Court for the Southern District of New York was packed with reporters as Apple and the Department of Justice finally faced off in the e-book price-fixing trial. Banned from bringing cell phones, laptops, and other electronic devices into the building, journalists were relegated to taking notes with pen and paper. The benches were hard and the over-air-conditioned rooms chilly.

Apple's legal counsel had a big hurdle to overcome. After reading the judge's pretrial comments saying that she believed the government would be able to prove the price-fixing, Apple's lawyers were concerned about their clients' ability to receive a fair trial.

"We respectfully and humbly ask this court to erase, to hit the delete button on any tentative view that might exist in the court's mind today," said Apple's lead attorney, Orin Snyder, in his opening statement. "We will demonstrate that Apple not only didn't conspire but should be applauded and not condemned for its . . . beneficial impact on the e-book market."

Almost immediately, the judge cut him off, saying that she only offered her pretrial views with the consent of both parties.

"This isn't a vote about whether I like Apple or anyone else does," Cote said. "The deck is not stacked against Apple unless the evidence is stacked against Apple."

In a nearly three-hour opening statement, Snyder argued that Apple had acted purely in its own interests, and that the pricing arrangement that the company had reached with each publisher was widely used in business and not illegal. Citing a U.S. Supreme Court case involving Monsanto, Snyder argued that Apple had legitimate reasons as a distributor to discuss prices with the publishers. Moreover, he asserted that Apple introduced healthy competition in a market that had been dominated until then by Amazon. He claimed that e-book pricing had fallen overall and the number of digital books available had increased since Apple launched its bookstore.

"Apple is going to trial because it did nothing wrong," declared Snyder. He accused the government of presenting evidence out of context and distorting the facts. "It's the first time in the history of antitrust laws that a new entrant coming into a concentrated market . . . is condemned," he said, calling the case "bizarre."

The government pressed the facts. Relying on 834 exhibits, including records of emails, telephone calls, and text messages between Apple and the five publishers that had also been named in the suit, the government vowed to prove that Apple had coordinated an old-fashioned price-fixing conspiracy by giving the publishers the leverage they needed to force Amazon to change its business model.

The specter of Steve Jobs loomed in this courtroom, just as it had in the San Jose trial. A key component of the government's case was a comment that Jobs had made to Walt Mossberg in an interview, that book prices would be the same across all of the retailers. Another piece of evidence included remarks that Jobs had made in Isaacson's book.

"Amazon screwed it up," Jobs had told the biographer. "It paid the wholesale price for some books, but started selling them below cost at $9.99. The publishers hated that—they thought it would trash their ability to sell hardcover books at $28. So before Apple even got on the scene, some booksellers were starting to withhold books from Amazon. So we told the publishers, 'We'll go to the agency model, where you set the price, and we get our 30%, and yes, the customer pays a little more, but that's what you want anyway.' But we also asked for a guarantee that if anybody else is selling the books cheaper than we are, then we can sell them at the lower price too. So they went to Amazon and said, 'You're going to sign an agency contract or we're not going to give you the books.' "

From the beginning, it was clear that the trial would come down to the testimony of Eddy Cue. As Jobs's deal maker and right-hand man for e-commerce and media, Cue had been the point person in the six weeks of negotiations with the publishers before the introduction of the iPad and the iBookstore on January 27, 2010. While the government sought to prove that Cue had played a central role in forcing Amazon and others to change their business models, Apple claimed that Cue had only reached the best deal that he could, without regard to the publishers' relationships with other retailers.

Cue took the stand on a Thursday morning in the second week. Shedding his usual jeans and button-down shirt, he appeared in a gray suit, white shirt, and a red tie. In the courtroom, as he answered the government's questions, Cue was so smooth that he almost seemed cheerful. When he disagreed, he did so calmly, saying, "No, that is not correct."

According to Cue, Apple had initially considered buying books at wholesale and selling them for a profit. But when the company found that Amazon was buying books at wholesale and selling them at a significant discount, Cue had to figure out another business model. He had seized on the agency model when two of the publishers suggested it. The publishers would benefit because they could set the prices, and Apple would be guaranteed a 30 percent cut of revenues no matter what.

In the course of the questioning, Cue readily admitted that e-book prices had risen after Apple opened its iBookstore and that he had met with all of the publishers and offered them the same terms. He even acknowledged that for a short time, Apple had sought to force publishers to impose the agency model on all of their retailers.

Cue explained that he backed off from the requirement, not because he realized the provision would be illegal but because there was no way for him to be sure that the agency deals that the publishers struck would be the same across the board. He was worried about the leverage that Amazon and Barnes & Noble would have from their physical book retail business. He also concluded that he had no ability to enforce an agency structure on Apple's competitors if the publishers couldn't persuade them to agree.

Though the judge had carefully remained neutral, she seemed to be coming around to Apple's perspective. The rules against horizontal price-fixing were clear. Companies in the same industry cannot collude to fix prices. But Apple was not a fellow publisher. It was a distributor. Its relationship to the publisher was vertical. Apple had made a compelling case that it was expected to negotiate the best price it could. In taking the actions that were construed by the government as price-fixing, Apple could have in fact been acting in its own interests. Apple may have exploited a situation to its benefit. But as a late entrant to the market, was it really capable of orchestrating a price-fixing scheme?

"I look forward to the summations," she said in the last moments of the trial. "To me, the issues have somewhat shifted during the course of the trial. . . . As you see it play out in the courtroom, things change a little bit."

Almost three weeks later, the judge handed down her decision. Apple, she said, was liable.

"Apple seized the moment and brilliantly played its hand," Judge Cote said. "It provided the Publisher Defendants with the vision, the format, the timetable, and the coordination that they needed to raise e-book prices."

Throughout her 160-page ruling, she was brutally critical of Cue's testimony. Contrary to some predictions that Cue had won the trial for Apple, the executive had failed to convince the judge of anything except his unreliability.

The judge made her disdain clear in a series of what one reporter called "acid footnotes." She pointed out Cue's inconsistencies and a host of improbabilities in his testimony. She called Cue a "savvy negotiator" and referred to one of his denials as "brazen."

"[R]egrettably," the judge concluded, "he was not credible."

Cote was just as unrelenting in the main body of her ruling. She referred to Apple's evidence as "not persuasive" and said "many of the trial's fact witnesses employed by Apple and the Publisher Defendants were less than forthcoming." She also found Jobs's exchanges with his biographer revealing.

"Jobs himself," she noted, "was frank in explaining how this scheme worked."

She pointed out that Apple itself did not try to deny that the publishers had engaged in price-fixing but chose instead to argue that it didn't know about it.

Cote deemed Apple's participation in the conspiracy "essential" in making sure that a critical mass of publishers would move forward with the same agency terms. She asserted that the pricing agreements between Apple and the publishers had "destroyed" competition rather than promoted it.

As the judge had mentioned in the pretrial hearing, much of the background in her ruling read as if she had prepared them in advance. But the last thirty-eight pages addressed Apple's defenses one by one. Cote disputed Apple's claim that it had acted independently in its own interests because the scheme had required the coordinated effort and conscious commitment of the publishers and Apple. She acknowledged that agency and most-favored-nation agreements were not illegal in and of themselves, but she added that did not mean a company could use those practices to restrain trade. In response to Apple's argument that the publishers would have withheld their books from the market if the company hadn't entered the market, the judge wrote that there was no evidence that the practice would have ever become widespread.

"From its very first meetings with the Publishers, Apple appealed to their desire to raise prices," she said. The judge expressed doubt that Apple had approached the initial meetings with publishers with no agenda, given Cue and his team's accomplished professionalism, the thoroughly reported challenges in the publishing industry, and the company's own short timeline to reach agreements with publishers. "One could ask why Apple has taken pains to argue that the mid-December meetings were simply a commercial listening tour," she said before answering her own question—any finding that the meetings were more than casual confirms Apple's intent to raise prices.

"The evidence is overwhelming that Apple and the Publisher Defendants' 'minds met' and they moved as one to achieve their conspiratorial objective."

In the aftermath of the ruling, Apple's defense team suggested that it had always known that its chances of winning were low, but they had hoped to raise enough questions for an inevitable appeal, arguably the more important phase of the dispute. The publishers have continued to deny that they conspired or engaged in any illegal or wrongful conduct.

The ruling was nevertheless devastating. Not only was Cote a well-respected judge and her decisions known to be difficult to reverse, but Apple's failed defense was also the same message that the company projected in its branding: The company worked for the greater good and was not bound by mortal concerns like profit. A part of Apple's narrative was that it wanted to change the reading experience and give consumers a better tool. And to a great extent, this was true. As Judge Cote put it, Apple had allowed cooks to learn how to make beef bourguignon from Julia Child. Children could run their fingers over a touchscreen while reading Winnie-the-Pooh. But those good intentions did not alter or excuse how Apple had worked with the publishers to squeeze Amazon. The company's inability to convince the judge otherwise signaled a breakdown in its image.

The deflections that had triggered the judge's disapproval were essentially identical to the tactics that Apple's public relations department employed with reporters—picking apart questions and answering them narrowly or not at all.

For years, this method of fending off inquiry had protected Apple's sacrosanct brand. During the patent trial with Samsung, the carefully rehearsed testimony of the executives had dazzled the jury. On Capitol Hill, Cook's precisely modulated mix of deference and defiance had allowed him to charm the senators. These tactics had not succeeded during Cook's public grilling at All Things Digital, where he had faced questions from insiders who were not satisfied with the company line. In the e-books case, Apple's evasions had proved even more disastrous. During the heat of the trial itself, Judge Cote seemed to fall under Apple's spell. But after she had time to dissect the testimony of Cue and the other witnesses, she had called the company out on its evasions. Again and again she struck at the heart of Apple's meticulously crafted image of faultless virtue.

Cook's debacle in the red chair was a warning. So was the judge's condemnation.

Apple was losing control of its narrative.

Manifesto

On a cloudy morning on June 10, 2013, Apple's faithful made their annual pilgrimage to San Francisco for the Worldwide Developers Conference. As always, the scene outside the Moscone Center crackled with anticipation. Police stood watch while the staff, identifiable by their bright red shirts and black pants, directed the flood of foot traffic. A half hour before Tim Cook's keynote, people rushed to join the line that snaked around the building.

The five-day conference had long served as the center stage for Apple's ambitions—the place where the company lowered its defenses every summer to unveil its latest wonders, to grant the disciples a glimpse into Apple's next chapter and excite them about the future. Apple engineers held sessions and mingled with the attendees to offer advice and exchange ideas about how developers could build software for iPhones, iPads, and Macs. Organized and scripted entirely by the company, the keynote that kicked off WWDC was a safe haven of possibility. Here no reporters were permitted to pummel execs with impertinent questions about market share or stock prices. No black-robed judges sat on high, wagging their fingers at the gulf between the company's saintly rhetoric and its predatory behavior. In this hall, the microphone was controlled by Apple, not a gaggle of U.S. senators who couldn't figure out how to update the apps on their iPhones.

If the company was going to regain control of its narrative, this was the moment.

Even among the company's most ardent fans—those pilgrims waiting for the doors to open—the innovation drought was raising eyebrows. It had been more than three years since the first iPad was introduced, six years since the original iPhone, and almost twelve since the iPod. Apple had only released new iterations of these devices since then, but they had been incremental advances, not bold new ventures, and some of the recent updates had been clouded by snafus and disasters. The faulty antenna, the comic mishaps of Siri, the maps debacle. Was the losing streak finally ending?

A block from the event hall, one of the believers pondered these things as he handed out steaming cups of coffee from the bed of a Radio Flyer wagon. Jordan Eskenazi worked for a Colorado start-up called Push IO. Along with others from the start-up, he had flown into San Francisco the day before. He had assembled the wagon that very morning just to have something to hold the coffee in. Along with the caffeine, he was passing out stickers. Push IO was also hosting a party that night for seven hundred developers. Eskenazi was excited about attending his second WWDC even if it required him to pull a toy wagon through the streets of San Francisco. But when the conversation turned to Apple's prospects, he grew slightly uncomfortable.

Eskenazi didn't want to sound negative. Push IO had been co-founded by a former Apple manager and existed because of the app boom that Apple had spawned.

"I'm not going to say anything bad." He paused, then added wistfully: "They could be doing better. It's sad that Jobs isn't here anymore."

Even on this street corner, the emperor's ghost hovered. In the span of a few minutes, Eskenazi referred to the fallen CEO but did not mention by name any of the lieutenants who had succeeded him. Nearly two years after his passing, they still disappeared in his shadow.

Eskenazi hoped aloud that Apple would show sparks of its old self. That's what everyone at the conference wanted, he said.

"With Jobs gone," he asked, "what is the new big thing going to be?"

Whether Apple's executive team wanted to admit it or not, they still had not found a way out of the genius trap. This was the question that defined them, hounded them, haunted their every decision. Now that their visionary leader had morphed into a global icon, his influence untouched by death, would they ever find the imagination and force of will to reignite the flame? Or was Apple on its way, as many suggested, to becoming just another company? One of the most successful companies in the world, certainly, but even so, a company that excelled primarily at making profits, not changing the world.

Down the street, the masses were filing into the convention hall, eager to hear the answer.

When Tim Cook and his team stepped into the spotlight, they bull-dozed over any suggestion that Apple had lost its touch.

Soaking up the wild cheers of the audience, they boasted again and again about their company's creativity and innovation. They quoted surveys on how much customers adored Apple, touted the many awards Apple had won, bragged about how their next genera-tion of products was better than ever, easier to use, more beautiful, more amazing, more great, more spectacular. Within a few minutes, they were already running out of fresh superlatives, but they kept going, recycling the same praises over and over like a blissed-out prayer.

They previewed new voices for Siri, male and female, and prom-ised that Siri had grown smarter, and that the maps app was new and improved, too. They made these vows with straight faces, avoiding any acknowledgment of the humiliations that had necessitated these corrections. Under these lights, amid the fervor of the crowd, there was only room for triumph. The executives put up a big slide showing that the App Store had now reached fifty billion downloads, and they played a video showing off the opening of the latest Apple Store, this one in a truly stunning old theater the company had renovated in Berlin. They announced that the new MacBook Air would have up to twelve hours of battery life, and that apps on the iPhone would now update automatically, and that the company would no longer name each new version of its desktop operating system after species of big cats, the most recent of which was Mountain Lion. The new versions would be named after inspiring places in California. First up would be Mavericks, after the legendary surfing beach. They cued up "Gimme Shelter" on iTunes Radio, their new streaming app, and they played the monster opening to Led Zeppelin's "Whole Lotta Love"—a real achievement given Zeppelin's historic resistance to allowing its catalog onto any streaming service. They checked the weather at the North Pole and brought out a pair of robotics experts who credited Apple's products for giving them the tools to design toy racing cars piloted with artificial intelligence. At one point, some of the toy cars crashed off the little track that had been rolled out for the demo, but apparently that was part of the fun. Either way, the toy car app was going on sale that day.

The biggest news of the morning came when Cook announced the latest version of the iOS operating system for the iPhone and iPad, grandly pronouncing the software to be "the biggest change to iOS since the introduction of iPhone." Demonstrating the new system, software chief Craig Federighi raved about the new look and feel of his own product.

"It's unbelievable," he said. "It's just gorgeous. From the typography on this lock screen to the vitality of the background, the animation, to the home screen with these icons. It looks so great. It just looks fantastic."

Appearing on a video, speaking solemnly as always in his British accent, Jony Ive declared that Apple saw the iOS7 as "defining an important new direction and in many ways, a beginning."

When Phil Schiller introduced a new cylindrical Mac Pro microcomputer, he struck a nationalistic tone, emphasizing that Apple was assembling the computer in the United States. Left unmentioned was the fact that many of the pieces inside the Mac Pro would still be manufactured overseas. Turning to the big screen, Schiller showed a video that visually linked the design with the curvature of Earth, leaving no doubt about how magnificent Apple considered the new product to be. When the video ended, Schiller nodded with satisfaction as he delivered what was clearly a scripted line.

"Can't innovate anymore, my ass," he said.

As the audience hooted and cheered, the camera crew shooting the session for the live stream zoomed in cheekily on Steve Wozniak, who, earlier in the year, had made public statements on his old company's diminishing creativity. This shot, too, seemed planned. To pull it off, the crew had to scout out Wozniak's seating location in advance and then wait for Schiller's money quote. The gotcha moment was similar to when the Oscars telecast showed the faces of nominees as they learned whether their names were in the envelope or not. Was it a flourish of choreographed revenge, a staged dose of instant karma aimed at embarrassing Wozniak?

"In the phone market they are a little bit lagging," Wozniak had told Germany's *Wirtschaftswoche* back in February. "There is no real major cost difference in the parts of the current Apple products and the competition. Samsung is a major competitor and the reason is because they are making great products." Though it was not printed, he had also suggested in the interview that Apple should incorporate some of Google's and Samsung's features to make their products even better.

Once again, Apple was revealing its true nature: The endless self-congratulation, the perpetual breeziness so obviously straining to respond to roiling doubts. And the blade behind the back.

To his credit, Wozniak did not give his attackers any satisfaction. When the camera closed in, his face betrayed only mild interest. One blogger, watching the live feed, thought Woz looked sleepy.

The keynote was peppered with other attacks, most of them expressed in coded allusions decipherable only to those who followed the company closely. The jabs were too fleeting to dampen the celebratory mood, and all were delivered with a smile. But the blade came out again and again as the speakers sliced away at the critics, at the doubters, at Android and Samsung and other rivals. Now that Scott Forstall had been banished from the empire, he too was considered fair game. The former chief of mobile software, long a fixture in the company's inner circle before his firing, was never named in the snarky asides. But the initiated understood that Forstall was being mocked for his love of weaving skeuomorphic elements into earlier versions of the iPhone and iPad—the iBooks app that looked like a bookshelf, the calendar app that looked like a desk calendar, the game center app with the digital version of a poker table's green felt. These touches were now deemed to be laughably cheesy. In the new iOS7 software, they would be deleted, just as Forstall had been.

Fathoming the motives behind the ridicule required the skills of a Kremlinologist. Why was Apple's leadership wasting its time tearing down someone they'd already shoved out into the cold? Why so much animus over fake green felt? Did they really believe that the world recoiled under the tyranny of skeuomorphism? The put-downs may have been a kind of chest thumping, intended to trumpet the alpha ascendancy of Cook and Ive, who had led the overhaul of the mobile software. Maybe they wanted to underscore that they had won and that Forstall had lost. Even more intriguing was the possibility that the real target might not have been Forstall, but their visionary founder. In years past, Jobs himself had advocated introducing the virtual bookshelf and the other skeuomorphic elements into the iPhone and iPad. Was this Apple's odd way of declaring independence from his legacy?

"There is a lot of subtle dissing of Jobs's design philosophy going on here," an editor noted during a live commentary on the ReadWrite tech site.

"Subtle?" wrote another. "I detect no subtlety."

If Jobs's successors had sought to ease his memory into the background, they failed. Watching Cook and his team that day, it was impossible not to recall the absolute mastery that Jobs had displayed so many times on this very stage. He too had indulged in his share of bullshit, but at least his bullshit had been mesmerizing.

Wisely, Apple kept this year's presentation moving, shifting from Cook to Federighi to Schiller, from videos to songs to slides to demos. Throughout the shuffling, the audience declared its approval.

"We love you!" someone shouted at Cook.

The developers' enthusiasm was understandable. To a large extent, their financial success depended on Apple's. But for others, watching the keynote from around the world as it streamed to the Web, it was clear that Apple had not found its next game changer. For all the bragging about their innovative prowess, the truth was that the company's newest offerings were hardly revolutionary. Many of the updates to both the Mac and iPhone operating systems were features that rivals already offered and that customers had been clamoring for Apple to adopt.

"This all seems . . . sorta marginal," wrote the editor from Read-Write, commenting on the Mavericks operating system.

"What exactly is original here?" wrote another, describing iOS7. "I am not sure I see anything really groundbreaking here. But hey, it sure looks pretty."

The latest iOS looked sleek and modern, but the design lacked distinction, a cross between Windows Phone, Android, and Palm's WebOS software. The black cylinder of the new Mac Pro was undeniably beautiful, even slightly dangerous-looking—like something Q might invent for James Bond. But the high-performance workstations would be expensive and were meant for professionals who needed server-level capacity and power. The majority of Apple's customers had no use for the thing. Rumors of iTunes Radio had generated buzz ahead of the formal announcement, but despite the snagging of Led Zeppelin, it was unclear how it would compete with Pandora, Spotify, and other streaming programs that had existed for years. Most disappointing of all, none of the products Apple had shown were going to be available before fall. And no one had said a word about Apple TV.

Cook closed the keynote with a reminder.

"Our goal at Apple," he said, "is to make amazing products that our customers love."

The line always played well, especially since it reinforced the impression that Apple was somehow nobler than its competitors. But ultimately the argument was meaningless. Didn't all companies seek to make something good? Did any CEO wake up in the morning, hoping to churn out products that customers disliked? For all the digs Apple had made at Samsung for its lack of originality, there was no doubt of Chairman Lee's attention to quality and reliability.

Still, Apple was fundamentally different from its rivals. Its immense success had been based on something unique—a record of achievement that went far beyond the ability to make good and reliable products. Until the death of Jobs, the company's fortunes had been made through its almost supernatural ability to create devices that were radically, irrevocably, inescapably new—machines so astonishing that they literally changed the daily lives of humans around the globe. Winning as the conference may have appeared to the audience, the entire event glossed over the fact that Apple seemed on the verge of forgetting how to do what it had done better than any other corporation. The high-tech evangelizing, the toys and the pretty lights, the insistence that the company was beginning anew, that its latest offerings were astonishing advances, that its wizards of design and engineering had retained their powers to remake reality—all of it was forced, overwrought, empty.

The final disappointment came at the end of the keynote, shining on the big screen at the front of the hall, when Cook previewed a new ad that would be the centerpiece of the company's next marketing campaign. The commercial wasn't about a particular product. It was intended as an assertion of the company's values, an attempt to redefine itself after the death of Jobs.

A manifesto.

"This is it," a man declared in a voice-over. "This is what matters."

As a spare melody joined the words, the screen surged with glimpses of people using Apple's products around the world. A young woman on a crowded New York subway, disappearing inside the music playing from her headphones. Elementary school children in a Japanese classroom, leaning over their iPads, their small fingers wiping across the screen, their faces lit with excitement.

"The experience of a product," the voice-over continued. "How will it make someone feel? Will it make life better? Does it deserve to exist?"

The scenes flowed on. A young couple hugging in the rain, with the woman holding out her iPhone to snap an image of them together. A man at a sushi bar, laughing at something on his iPad mini. An ocean of people dancing at a rock concert, holding up the lighted screens of their iPhones in front of the stage.

"We spend a lot of time on a few great things, until every idea we touch enhances each life it touches."

An older couple at a banquet, beaming and laughing as photos of their life together are projected from a MacBook onto a screen. A teenage girl in her room, the walls glowing with Christmas lights as she checks her iPhone.

"You may rarely look at it, but you'll always feel it. This is our signature. And it means everything."

As the screen went black, a simple sentence appeared in white letters.

"Designed by Apple in California."

The words were pretty. The images were moving. But the ad said nothing beyond what Apple had been repeating for years. The most striking touch was the designed-in-California slogan at the end, the tagline that would anchor all of the ads in the campaign. To some, the phrase was reminiscent of the made-in-America slogans and songs that General Motors and Chrysler had deployed in the past when confronted with competition from Toyota and Nissan. Was Apple slamming Samsung, reminding American consumers which company was based in Seoul? Or was Apple simply hoping to bank on California's residual cool?

One possibility veered dangerously close to jingoism. The other was just lame.

When the "Our Signature" ads began to air later that day, the response was dismal. Ace Metrix, a consulting firm that gauges the effectiveness of TV ads through viewer surveys, gave the new ad the lowest score of twenty-six Apple TV ads in the past year. *Adweek* published a searing analysis titled "At a crucial time of transition, has the company simply lost its voice?"

The author, Tim Nudd, reminded his readers that Cook had talked about how much the words in the ads meant to everyone at Apple. "Here, though, they fail to do the trick," Nudd wrote. "They aim for poetry in the classic Apple style. But maybe it really isn't the same company after all."

Those who liked the ads compared them to the "Think Different" spots, the first campaign that Jobs had worked on when he returned to Apple. The circumstances were certainly similar. Apple had no new product, but the company needed to convey what it stood for to grab the attention of consumers and give the employees something to rally behind during a time of turbulence. Yet there was a fundamental difference in tone between "Think Different" and the latest campaign. "Think Different" had been about Mac users. It overlaid Apple and

Jobs's crazy image, making the point that people who do great things are often misunderstood and labeled as troublemakers. At the same time, it celebrated trailblazers who made a difference in the world and drew a connection between them and Apple's customers.

The new manifesto was about what Apple stood for, not what the customers stood for. Ultimately, it was another way of insisting that Apple was still great.

"I don't care for manifestos that try so hard to be a manifesto," said Ken Segall, a veteran copywriter who had worked closely with Jobs on "Think Different" and other campaigns.

In the past, Segall pointed out, Apple's ad campaigns had conveyed their messages indirectly through imaginative, surprising images and words that made people identify their best selves with the company's aspirations. The ads didn't tell customers to buy Apple devices by claiming that the products were hip. That almost never worked in advertising. Instead, the ads allowed viewers to judge Apple's appeal on their own.

Cool people rarely bragged about their own coolness or even spoke of it aloud. Great companies did not need to insist on their greatness.

In the months that followed, as the new manifesto wore thinner with each airing, the temptation grew to think back to that day two years before, when the masses had gathered at One Infinite Loop to remember Jobs. Sometimes, the service seemed long ago. Other times, it seemed like yesterday, each detail vivid in the mind's eye.

The giant banners adorned with their emperor's face, watching unblinking from above. His lieutenants, standing on that stage, trying so hard to act as though they knew what to do next. And then that unmistakable voice, washing over them all.

Here's to the crazy ones. The misfits. The rebels. The troublemakers. The round pegs in the square holes. The ones who see things differently.

He'd been the craziest of them all. The most arrogant. Matchless in his cunning. Both brazen and terrifying.

They're not fond of rules. And they have no respect for the status quo.

The truest of the true believers, he had seen straight into the heart of their creations. He knew what each of them wanted, what they feared. The minefields waiting down the road.

You can quote them, disagree with them, glorify or vilify them. About the only thing you can't do is ignore them.

Forgetting him was like trying to forget the sun. He still reigned over every hour of every day. That was his blessing, and their curse.

Epilogue

NOVEMBER 2013
Less than two years after the death of Steve Jobs, the empire he left behind at the peak of its influence now teeters at the edge of a reckoning. The more his successors insist that the company is stronger than ever, the less the public believes them.

Despite the best efforts of Apple's propaganda machine, the response to the latest wave of upgrades and incremental advances has been tepid. In July, the company reported its second consecutive decline in profits and flat revenues for its fiscal third quarter. iPad sales had fallen 14 percent, and iMacs 7 percent. iPhones sold better than expected, but consumers were choosing the cheaper, older models, which meant that Apple was making less money from each sale. In early August, Apple scored a victory when the Obama administration vetoed the U.S. International Trade Commission's ban on the import and sale of some of the company's older iPhones and iPads. The ITC had found the devices to infringe on a Samsung patent, but the federal government stepped in because it was concerned about its impact on competition. Since most of the devices in question were no longer even on sale, the win was largely symbolic. The company did not triumph through the righteousness of its position. It won, sort of, because the administration wasn't ready to publicly abandon a quintessentially American company.

From the start, Apple embodied this country's idea of exceptionalism. Steve Jobs and his team trumpeted the notion that their company was fundamentally different—a moral enterprise, imbued with a desire to remake the world, destined for greatness with reserves of ingenuity and strength that render it immune to the failings of lesser rivals. In the halls of Congress and in the Oval Office, that presumption is still being argued for America. And in the winding corridors at One Infinite Loop, the concept still holds tremendous sway. Tim Cook and his senior executives clearly still believe that Apple is ethically and creatively superior to other companies.

The truth is, Apple used to be exceptional. Not necessarily in its behavior, which was often predatory. But certainly in its ability to inspire. Those days are waning. Outside the echo chamber of Apple's headquarters, the notion of the company's exceptionalism has been shattered. The revelations of worker conditions at Foxconn, of Apple's strained relationship with its suppliers, and of the company's elaborate attempts to avoid billions in American taxes have made it impossible to pretend that the company exists on a higher moral plane. In the e-books case, Judge Cote's ruling dismantled Apple's credibility.

In early September, Cote meted out her punishment for Apple's transgressions in the e-book conspiracy. In her final ruling, the judge restricted the company's ability to enter into new agreements with publishers. The injunction wasn't as restrictive as the government had requested, but the judge ordered an external monitor—a watchdog—to ensure that the company complied with antitrust laws. The decision boiled down to a single conclusion: Apple cannot be trusted.

"Apple has been given several opportunities to demonstrate to this court that it has taken the lessons of this litigation seriously," Cote had said in a hearing. "I am disappointed to say that it has not taken advantage of those opportunities." Apple appealed the ruling in October.

In other courtrooms, from California to Germany, the patent wars continue with no end in sight. Speculation ebbs and flows about a settlement that would end Apple and Samsung's global conflict, but so far the battalions of lawyers march on. Much of what the jury decided in San Jose in the summer of 2012 has already been watered down. Both companies are now gearing up for a partial do-over trial later this month. Another trial, covering a different set of patents in contention between the two tech giants, is set for 2014. At this point, Apple's unceasing pursuit of legal challenges is at best quixotic.

To some, the insistence on pursuing legal remedies points to a deep miscalculation. In a global economy that is rapidly evolving ahead of laws and jurisdictions, the market reigns supreme more than ever. Will the company go after every new rival that springs up around the world, selling cheaper phones that look like the iPhone? In this new century, innovation flows like water.

Whatever Apple's future, China looms. The company's hopes for increasing its profits are still pinned on selling more of its devices in the rapidly expanding Chinese market. For years, Apple has been negotiating an iPhone deal with China Mobile, the world's largest mobile phone company in terms of its subscribers. But they have struggled to reach an agreement. The new reality is that Apple needs China Mobile more than China Mobile needs Apple. In a country with more than a billion people, the Apple brand is already losing its sheen. With earnings far below that of American workers, Chinese consumers are veering toward cheaper Android smartphones—products made not just by Samsung and HTC but by up-and-coming Chinese companies such as Lenovo, Huawei, and Beijing Xiaomi Technology.

Xiaomi's rise illustrates the futility of crushing imitation. Xiaomi sells iPhone-like devices at a fraction of Apple's prices. Lei Jun, the CEO, has long held up Jobs as his role model. In the past, Lei emulated the fallen CEO so thoroughly that he often appeared onstage in jeans and a black shirt. Today, Lei's admiration has faded. Though Xiaomi was known as the "Apple of the East," Lei now compares his company to Amazon. In a recent interview with CNN, he said Apple didn't seem to care about what users wanted.

All around the world, Apple's competition keeps gaining. According to IDC, almost 80 percent of smartphones shipped in the second quarter of 2013 ran on the Android operating system while Apple's share declined more than three points to 13.2 percent. Making matters worse, there are signs that the industry may be headed for a slowdown in smartphone sales now that most customers who want one have already bought one.

Samsung's market share had also slipped over the summer, but in the second quarter of 2013, it still held a 30.4 percent share according to IDC. Its quarterly profit rose 50 percent to 7.77 trillion won, or $6.9 billion, during the same period.

The bigger concern for Samsung is arguably no longer Apple. It's the up-and-coming Chinese companies, already plotting to dethrone its older rivals in Seoul and Cupertino.

As fall approached, the pressure kept building on Apple to unveil the kind of innovations the world was clamoring to see. In September, Apple launched two new phones, the iPhone 5s and a plastic iPhone 5c that was $100 cheaper. The 5s came with a fingerprint sensor and a faster chip. It was also available in gold in addition to silver and "space gray." The 5c, effectively last year's iPhone with a larger battery, came in five colors. Both devices included the latest version of the iPhone operating system.

Apart from the excitement over the gold model, the public's reaction was lukewarm. Reviewers again pointed out that the upgrades were hardly revolutionary. Many noted that the price of the 5c, while less expensive, was hardly cheap. With a two-year contract, U.S. carriers subsidized the phones so consumers could purchase them for as little as $99, but in China, the 5c would cost $733 without a subsidy. Longtime Apple watchers wondered if the new lineup was an attempt to distract users by dazzling them with new colors. Apple had done the same with their iPods as the market approached saturation. Apple's shares fell 2.3 percent that day and continued to decline in the following days.

The reviews on the aesthetically overhauled iOS 7 software were also mixed. Many liked the simpler, more modern interface with bright colors and flat icons. The high school teacher John Keitz was thrilled to find that Siri now understood him even when he issued a command in a loud kitchen. Others deplored the fact that the operating system crashed apps and navigating the phone was less intuitive.

When the new phones went on sale ten days after the unveiling, Apple boasted that first weekend sales hit a record nine million devices, 80 percent more than the previous year, but analysts were quick to point out that Apple was selling two phones this year rather than one as in the past, and it had released the phones in eleven markets, three more than the previous year. China, part of the initial launch for the first time ever, was estimated to have contributed more than two million of the sales. Early signs were particularly foreboding for the 5c: The model was widely available ten days after Apple began accepting pre-orders.

In the weeks since then, it has become apparent that the 5c is not selling. Shortly after the new iPhones went on sale, Best Buy and Walmart halved their 5c price with a two-year contract, but it had little effect. A couple of weeks later, the media reported that Apple had cut its 5c orders. That action has reverberated down through the supply chain once again.

The spoof publication, the *Onion*, practically predicted the public's reception in a piece after the launch.

"Apple Unveils Panicked Man With No Ideas," said the headline above a photo of Tim Cook. The article pointed to the CEO's "artificial excitement," "quavering voice," and "near total lack of inspiration."

A decline was inevitable.

The story follows an archetypal pattern—a pattern familiar in both history and myth. A struggling empire, on the brink of dissolution, recalls one of its founders from exile and casts him as a savior. The ruler, ruthless and cunning as Odysseus, gathers the faithful and emboldens them to take startling risks that allow the empire to reach even greater heights than before. Amid the celebrations, the emperor grows sick. Knowing that he is the living embodiment of his kingdom's fortunes, he tries to hide his illness until he is finally forced to accept that he is not immortal. Left to carry on in his name, the emperor's lieutenants fall prey to complacency and confusion, lapsing into disarray and paralysis. Bound to the way things have always been done, these new leaders become less flexible and ignore the warning signs. Their emperor is gone, but ever present. Though they are still at war with enemy armies, these lieutenants cannot find their own way forward. They are tired. They are uncertain. The well of ingenuity has run dry.

The theories of Clayton Christensen and Gautam Mukunda explain it all so clearly. Apple's story, as they have noted, has long unfolded in defiance of business physics. With each new triumph, the company rose higher and higher. But sooner or later, gravity always wins.

On Apple's campus in Cupertino, morale has languished, and employees are quitting. Apple has always asked a great deal of its faithful. But in the past, they accepted the long hours, the ceaseless pressure, and the verbal tirades because they felt that they were working for a company always reaching for something higher. When they saw Jobs on stage, unveiling the latest wonder that all of them had joined together to create, they had felt fulfilled. After the release of the first iPad, though, it became harder for many of them to see the point of it all. Each wave of new products felt more incremental, less revolutionary, less wondrous.

In the past two years, as it has become clearer that something fundamental has changed at Apple, the pace of the resignations has increased. Some employees wait, biding their time until their stock vests. Others just leave. In some parts of Apple, farewell gatherings have become a weekly ritual. A new phrase has been coined to describe some of the defections: G2G, or Go To Google.

In October, the company launched two new iPads—a thinner, lighter full-size tablet called the iPad Air and an updated, faster iPad Mini with a sharper display. It also said that it would make free its iWork productivity suite and iLife photo, movie, and music-making apps, which had previously sold for $4.99 to $9.99 each. Cook called the changes "the biggest iPad announcement ever, by a large margin," but the boasts rang hollow. iPad sales had been falling as Samsung and Amazon's cheaper devices eroded the company's once dominant share. Gartner research firm projects that Apple will split the market with the Android camp almost equally in 2013.

After the event, Marco Arment, the creator of Instapaper, noted on his blog that most of the presenting executives didn't even seem excited about their own announcements.

"At best," he wrote, "the presentation felt uptight."

Shortly after that, Apple reported its third consecutive quarterly decline in profits and just a four percent increase in revenues. Both iPhone and iPad sales were below analysts' projections.

When the iPad Air went on sale, Walmart, Staples, and Target cut the price of the entry-level model by $20 from the get-go.

Despite the launch of the new iPhones and iPads, the hunger for something fresh, some beautiful breakthrough that will turn everything around, continues. Apple's inventors are reportedly at work on new products—a connected watch device, the long-rumored television. Taking a cue from Samsung, the company has also been experimenting with phones of different sizes.

It's not too late for Apple to dazzle the world again. The magic can still be resurrected. Some reports suggest that the company might unveil the watch or the TV sometime in 2014. But at this point, the company has grown so huge that it would need to sell millions of these new products to make a significant enough impact on overall sales and profits to be considered successful.

The debate continues over what would have happened if Jobs were still here to run the place, revving imaginations and terrifying underlings, speeding through life in a car with no license plate—rewriting the rules every day of what is possible for him and for Apple. Many of the problems that have plagued his successors were already taking root before his death. Whatever answers Jobs might have offered, had he lived, it's unlikely that he would have been second-guessed. He might even have found a way to convince people for a while longer that Apple was as great as ever.

Without him, everything changed. The dilemmas multiply and deepen. Solutions slip further out of reach.

It's important to remember that Jobs handpicked his successor. He trained Tim Cook, indoctrinated him, subjected him to trial by fire, and spent years evaluating his capabilities. Jobs chose Cook knowing full well that he was not a visionary, that he had no history as an innovator, that he was best known for his devotion to spreadsheets. Jobs didn't choose Jonathan Ive to lead Apple. He chose the Attila the Hun of Inventory. The question is, why? Did he think that a numbers man was best for the storms gathering at the horizon? Or did he want to make sure that his vision would not be replaced by someone else's?

Since Cook took over, he has often explained that his mission is not to emulate what Jobs would have done, but to always do what is best for Apple. At the same time, he has repeatedly insisted that nothing has changed at Apple, even as the world has changed around him. It's unclear if he sees the contradiction.

Watching Cook in public—taking the stage for another launch, sitting in the red chair where his boss once ruled—is to behold a man laboring at an impossible task. He's smart. He's engaged. He believes. But the words he utters come out flat and slightly off-key. There is no spark. No fire.

Even some of Apple's biggest supporters have noted the decline. Late this summer, in an interview with *60 Minutes,* Larry Ellison talked about what Apple was like when Jobs was pushed out in the eighties, and what it was like after he came back.

"We conducted the experiment," said Oracle's CEO, a former Apple board member and one of Jobs's closest friends. "I mean it's been done. We saw Apple with Steve Jobs. We saw Apple without Steve Jobs. We saw Apple with Steve Jobs."

Now the pattern was repeating itself. Only this time, Jobs had left One Infinite Loop forever.

"Okay, I'll say it publicly," Ellison admitted. "I don't see how they can . . . they will not be nearly so successful because he's gone."

The newsstand at the airport tells the story a different way. Two magazines, on sale at the same time, with two competing covers, catch your eye. The issue of *Time* asks the following question:

CAN GOOGLE SOLVE DEATH?
The search giant is launching a venture to extend the human life span. That would be crazy—if it weren't Google.

BloombergBusinessweek, displayed on the shelf below, shows Cook laughing triumphantly with two of his lieutenants. The headline reads:

WHAT, US WORRY?
Exclusive: Apple's Craig Federighi, Tim Cook, and Jony Ive have never been more certain they're right

The contrast is striking. *Time*'s story details how Google is creating a new company designed to search for medical advances that will help our species face down mortality. The piece describes other Google projects that, if they even reach fruition, will require decades of work—enterprises aimed not at pumping up profits but taking the long view to confront challenges that have vexed humanity since the dawn of time. Inevitably, the article draws comparisons to Google's famous rival.

"Last week Apple announced a gold iPhone; what did you do this week, Google? Oh, we founded a company that might one day defeat death itself."

The *BloombergBusinessweek* piece is a pep talk about Apple's future. The writer notes the predictions of the company's demise, then lets Cook and his team explain why the doubters are wrong. Cook repeats the familiar arguments that everything is fine, that he's not worried, that he and his team have everything under control. On Apple's shrinking smartphone market share, Cook points out that much of the decline is due to the proliferation of so many cheap phones.

"There's always a large junk part of the market," he says. "We're not in the junk business."

The article rehashes the company line that Cook has been repeating now in every interview. There is no hint of a long view, of anything happening at Apple beyond the latest product or the next quarterly earnings report.

They're not reinventing the world. They're circling the wagons.

Glance back at the cover. Cook and Federighi and Ive look so insistently happy, so determined to project confidence.

What did they think when they saw the headline? Did it occur to them that "What, us worry?" is a steal from *Mad* magazine? Do they remember seeing Alfred E. Neuman's grinning jughead on the cover next to his ironically nonchalant slogan, "What, me worry?"

Yes, the new emperor and his team are laughing. But do they get the joke?

Acknowledgments

At the start of this project, I was warned about the solitary nature of writing a book. My experience has been anything but.

My deepest gratitude to the nearly two hundred former and current Apple employees, business partners, and others in the Apple world who spoke to me. They could not have been more generous with their assistance. The list is long, and I agreed to protect most of their identities, but I'd like to acknowledge three—Fred Anderson, Jon Rubinstein, and Avie Tevanian. They did not necessarily agree with my conclusions, but they verified background where they could and challenged my thinking in other areas. My respect for them has only grown.

In tackling this enormously complex story, I consulted experts to help me understand the intricacies of various subjects. In addition to those who are quoted in this book, they include Dr. William Chapman, Klaus Klemp, Robert Robins, Ethan Bernstein, Simon Yang, Colley Hwang, James McQuivey, Gary Allen, Jacquelyn Wilson, and Toni Sacconaghi. Foxconn spokesman Louis Woo was most gracious in his help as he chased tedious details, down to the menu for the Stars of Foxconn banquet. In my reporting travels, Ms. Evelyn Lowery and *WKRG*'s Debbie Williams were my guides in Alabama and my crackerjack assistant Tian Yuan (Violet) in Hong Kong and Shenzhen. PixelQi CEO John Ryan and Nikkei BP contract manufacturing expert Tomohiro Otsuki introduced me to their contacts and made me feel at home in Taipei. BlueRun Ventures' Kwan Yoon helped me navigate Seoul.

One of the most rewarding aspects of working on this book was encountering the extraordinary generosity of fellow journalists. Walter Isaacson provided me with his support and shared background information that didn't make it into his Steve Jobs biography. Joe Nocera at the *New York Times* not only gave me details about his famous conversation with Jobs, he let me tell the story first. My friend and *Wired* writer Fred Vogelstein trusted me with an early copy of his own book in addition to fishing me out of my darkest moments. Poornima Gupta at Reuters, Ina Fried at *All Things Digital*, and Martyn Williams at *IDG* helped me with details on the Apple-Samsung patent trial. *Fortune*'s Philip Elmer-DeWitt shared his reporting on the e-book trial and *BloombergBusinessweek*'s Josh Tyrangiel gave me background on his interview with Tim Cook. Matthias Hohensee of *Wirtschaftswoche* sent me the original transcript of his interview with Steve Wozniak, mentioned in chapter twenty. Across the Pacific, Diamond Weekly's Naoyoshi Goto and Jun Morikawa shared their reporting on Apple's supply chain in Japan. Nikkei BP Publishing's Hiromi Nakagawa also helped with my research on Japanese suppliers.

There are many others who helped me with this book but asked to not be named because they weren't authorized to speak to me or because of their ties to Apple, Samsung, and Foxconn. You know who you are. Thank you.

My thanks also to everyone I've worked with at the *Wall Street Journal*. News Corp. CEO and former *WSJ* managing editor Robert Thomson and deputy editor-in-chief Matt Murray supported my initial idea for a book and granted me a leave of absence that allowed me to set off on this path. During my three years covering Apple for the paper, my editors in the San Francisco bureau—Steve Yoder, Pui-Wing Tam, and Don Clark—pushed me to be a better reporter than I had ever imagined I could be. Alix Freedman and Mike Williams made sure my page-one stories were airtight. My colleagues around the world helped me break the stories that led to this book and provided me with their continued help and friendship. My gratitude especially goes to Ian Sherr, but also to Geoffrey A. Fowler, Joann S. Lublin, Thomas Catan, Nick Wingfield, Jason Dean, Loretta Chao, Ting-I Tsai, Evan Ramstad, Ethan Smith, and Amir Efrati. Thank you also to Walt Mossberg and Kara Swisher at *All Things Digital* for their encouragement, advice, and help.

Of course, none of this would have been possible without a team of the finest reporting assistants. In addition to Violet, special thanks to Natalie Jones in the United States who has been with me since the very beginning. Bobby Tung in Taiwan, *WSJ* Seoul assistant MinSun Lee, the *Korea Times*'s Kim Yoo-Chul, and *Korea JoongAng Daily*'s Kim Hyung-eun also provided invaluable assistance. Laila Kearney helped fact-check.

If it weren't for my *WSJ* colleague Jim Carlton, who wrote a fabulous book of his own on Apple over fifteen years ago, I would have never met the best agent anyone could ever have. Peter Ginsberg has been an anchor, and I couldn't be more grateful for his steadfast guidance and encouragement. He led me to Jonathan Burnham and Hollis Heimbouch at HarperCollins. Both have been unflagging in their support and enthusiasm for the book. It's rare to find an editor who not only understands an author's vision, but also cares about the nuances of what she is trying to say. I'm incredibly lucky to have Hollis and her extraordinary team in my corner. They include her associate editor Colleen Lawrie, copy editor Tom Pitoniak, jacket designer Milan Bozic, marketing manager Stephanie Selah, and publicist Steven Boriack.

Writing a four-hundred-page book is a daunting undertaking, particularly for a newspaper reporter who has never written more than three thousand words. I owe a huge debt of thanks to Tom French, who took me under his wing and gave me a master's education in narrative writing. Thanks to him, I now use words like "scaffolding" and scrutinize structure in every movie and book I come across. Tom O'Keefe read each chapter multiple times with the careful eye of an AP English teacher, Kelley Benham French with the eye of a newspaper editor, and Tim Schaaff from the perspective of a former Apple executive. Randall Farmer gave the manuscript a geek's read. Each was indispensable.

I would be remiss if I didn't thank the San Francisco Writers' Grotto for welcoming me into their fold. They gave me not only an office but the talented community of five-dozen fiction and nonfiction writers who also offered me friendship, advice, and moral support. I am proud to work among them.

There is nothing like writing a book to make one so completely self-absorbed, yet my friends and family stuck by me through the ups and downs. I dare not name my friends for fear of forgetting someone, but I'd like to thank both sets of my parents, Don and Fumiyo Iwatani and Dennis and Kathleen Kane for believing in me; my chef sister Yuki for keeping me nourished; and Mayu, Ryan, and Sue for their support and love.

Finally, I want to thank my husband, Patrick. Words cannot express how much his encouragement has meant throughout my career. He is my toughest critic, my biggest champion, and my best friend. It's because of him that I can do what I do.

Notes

Prologue: I Used to Rule the World

1.   **That Wednesday:** A Celebration of Steve's Life, Apple video, October 19, 2011, http://events.apple.com.edgesuite.net/10oiuhfvojb23/event/index.html; "Apple Celebrates Steve Jobs at Company Headquarters," CNET video, October 19, 2011 http://cnettv.cnet.com/apple-celebrates-steve-jobs-company-headquarters/9742-1_53-50113464.html; "Recording of Steve Jobs Is Highlight of Memorial Service at Cupertino Campus," *San Jose Mercury News*, October 19, 2011; the account is also partially based on interviews with several Apple employees who were present at the service or had watched it remotely.

Chapter 1: The Disappearing Visionary

7.   **For a company:** Author's personal observations; interviews with former executives and managers, including one person who was part of the campus development team when it was being built. Apple considers only members of the executive team to be executives, but for the purposes of this book, an executive is anyone who is a vice president or higher.

7.   **One Infinite Loop:** The street was named after a programming concept in an employee contest. The runner-up was Floppy Drive, but it was nixed because it sounded outdated.

9.   **One morning that summer:** Interviews with two people who were present; Michael Sippey, "Two Minutes With Steve," Sippey.com, October 6, 2011, http://www.sippey.com/2011/10/two-minutes-with-steve.html.

10.   **Only fifty-three, Apple's savior:** Jim Carlton, *Apple: The Inside Story of Intrigue, Egomania, and Business Blunders* (New York: Random House, 1997); "Steve Jobs Demos Apple Macintosh, 1984," YouTube video posted by tranquileyedotnet on November 7, 2006, http://www.youtube.com/watch?v=G0FtgZNOD44; "Apple—1984" YouTube video posted by antisubliminal on June 19, 2006, http://www.youtube.com/watch?v=R706isyDrqI.

11.   **"Do you really want to sell sugar water:** There are a few versions of this quote that are each slightly different. Some have Jobs saying "sugared water." A few have quoted Jobs as saying, "Do you really want to sell sugar water your whole life?" In September 2011, Sculley told the *Triangle Business Journal* that Jobs said, "Do you really want to sell sugar water or do you want to come with me and change the world?," leaving out "for the rest of your life." While none of these materially change the intention of what Jobs had said, the author went with the most widely used version.

12.   **Amelio was in way over his head:** Interviews with Gil Amelio, Edgar S. Woolard, and other executives, engineers, and employees at the time; Jim Carlton, "Thinking Different: At Apple a Fiery Jobs Often Makes Headway and Sometimes a Mess," *Wall Street Journal*, April 14, 1998.

12.   **After the board begged him to return:** Interviews with Apple executives at the

time; Alan Deutschman, *The Second Coming of Steve Jobs* (New York: Broadway Books, 2000), 255; Carlton, *Apple*, 433.

15.   **But in October 2003:** Interviews with Apple executives and employees at the time; Walter Isaacson, *Steve Jobs* (New York: Simon & Schuster, 2011), 453, 476; Peter Burrows, "Apple's Cancer Scare," *BloombergBusinessweek*, August 1, 2004, http://www.businessweek.com/stories/2004-08-01/apples-cancer-scare; John Markoff, "Talking of Chief's Health Weighs on Apple's Share Price," *New York Times*, July 23, 2008.

17.   **By 7 a.m. on the first day:** Based on an account by Nick Wingfield, former *Wall Street Journal* Apple beat reporter: Ryan Block, "Steve Jobs Keynote Live from WWDC 2008," Engadget, June 9, 2008; "WWDC 2008 Keynote Address," Apple Keynotes podcast, June 9, 2008; Nicholas Carlson, "Apple CEO Steve Jobs Looks Dangerously Thin," Gawker, June 9, 2008; Henry Blodget, "Apple (AAPL) Weakness and Steve Jobs' Scare Reveal Need for a Better Apple Plan," Business Insider, June 13, 2008, http://www.businessinsider.com/2008/5/apple-aapl-crushed-again-on-concerns-about-steve-jobs-health/.

Chapter 2: Reality Distortion

20.   **One Thursday afternoon:** Interview with Joe Nocera; Joe Nocera, "Apple's Culture of Secrecy," *New York Times*, July 26, 2008; Peter Elkind, "The Trouble with Steve Jobs," *Fortune*, March 5, 2008; author's interview with a person with firsthand knowledge of Apple contractor firing.

23.   **Once the idea:** Ryan Tate, "Steve Jobs' Obituary, As Run by Bloomberg," Gawker, August 27, 2008.

23.   **At a media event:** "Apple Special Event October 2008," Apple Keynotes podcast, October 14, 2008; author's own recollection of the event.

24.   **Several weeks later, it became:** "Apple Announces Its Last Year at Macworld," Apple press release, December 16, 2008; Yukari Iwatani Kane, "Apple CEO Will Skip Macworld Trade Show," *Wall Street Journal*, December 17, 2008; Yukari Iwatani Kane and Nick Wingfield, "Apple Shares Slump Amid CEO Worries," *Wall Street Journal*, December 18, 2008.

24.   **The truth, no matter:** Isaacson, *Steve Jobs*, 480–81; "Letter from Apple CEO Steve Jobs," Apple press release, January 5, 2009; "Apple Media Advisory," Apple press release, January 14, 2009; author's interview with Walter Isaacson as well as former Apple executives and employees who had firsthand knowledge of the situation. William C. Chapman, MD, Washington University in St. Louis transplantation chief, helped provide medical context. Robert Robins, a noted expert on disabled leaders, and Ethan Bernstein, a Harvard Business School professor specializing in leadership and organizational behavior, were consulted about corporate governance laws and CEO health disclosures.

28.   **The split in opinions:** Isaacson, *Steve Jobs*, 481–82; author's interview with a former Apple executive with firsthand knowledge of the events described; reporting associated with "On Apple's Board, Fewer Independent Voices," *Wall Street Journal*, March 24, 2010.

31.   **Aside from the occasional gossip:** Reporting associated with author's article "Jobs Maintains Grip at Apple," *Wall Street Journal*, April 11, 2009; interview with a former Apple executive with firsthand knowledge of the events described. The incident between the *Wall Street Journal* and the Apple spokesman at the end of the section refers to an exchange that the author had while she covered the company for the newspaper.

32.   **When Jobs went on medical leave:** Reporting associated with author's article with Joann S. Lublin, "Jobs Had Liver Transplant," *Wall Street Journal*, June 20, 2009;

Alexander Haislip, "Is Steve Jobs Moving to Memphis?," PEHUB, April 15, 2009; author's interview with Jim Gilliland, Jobs's neighbor in Memphis; Brian Caulfield, "What Are You Saying Now About Jobs?," Forbes.com, June 24, 2009; Yukari Iwatani Kane and Joann S. Lublin, "Apple Mum on Jobs's Treatment, Diagnosis," *Wall Street Journal*, June 30, 2009.

34.   **Jobs was motivated:** Interview with Walt Mossberg.

Chapter 3: Vertical

36.   **In January 2009:** "Apple Inc. F1Q09 Earnings Call Transcript," Seeking Alpha, http://seekingalpha.com/article/115797-apple-inc-f1q09-qtr-end-12-27-08-earnings-call-transcript.

37.   **Jobs was irate:** Interviews with sources familiar with the situation, including people with firsthand knowledge; "Apple Reports Second Quarter Results," Apple, April 22, 2009; reporting associated with author's article with Joann S. Lublin, "Absent Jobs, Cook Emerges as Key to Apple's Core," *Wall Street Journal*, June 23, 2009; Isaacson, *Steve Jobs*, 491.

39.   **After redefining computers:** Interviews with sources familiar with the situation, including people with firsthand knowledge; reporting associated with author's article "Jobs, Back at Apple, Focuses on New Tablet," *Wall Street Journal*, August 25, 2009; Isaacson, *Steve Jobs*, 494; "Apple Announces iPad," Apple Keynotes podcast, January 27, 2010; "The Book of Jobs," *Economist*, January 28, 2010.

41.   **Jobs was a master evangelist:** Isaacson, *Steve Jobs*, 505; author's firsthand reporting of *Wall Street Journal* editorial meeting; author's interview with sources familiar with the situation including a person with firsthand knowledge about the News Corporation discussions.

43.   **"Steve, how are you feeling":** This question was asked by the author in the meeting.

43.   **When Apple released the iPad:** Michael Arrington, "The Unauthorized TechCrunch iPad Review," TechCrunch, April 2, 2010, http://techcrunch.com/2010/04/02/the-unauthorized-techcrunch-ipad-review/; reporting associated with author's article with Ben Worthen, "Steve Jobs Escalates Fight with Adobe," *Wall Street Journal*, April 27, 2010; John C. Abell, "Google's 'Don't Be Evil' Mantra Is 'Bullshit', Adobe Is Lazy," Wired.com, January 30, 2010, http://www.wired.com/business/2010/01/googles-dont-be-evil-mantra-is-bullshit-adobe-is-lazy-apples-steve-jobs/; Steve Jobs, "Thoughts on Flash," Apple website, http://www.apple.com/hotnews/thoughts-on-flash/.

45.   **Apple's App Store:** Reporting associated with author's article with Thomas Catan, "Apple Draws Scrutiny from Regulators," *Wall Street Journal*, May 4, 2010; reporting associated with author's article with Jeffrey A. Trachtenberg, "Texas Questions E-book Publishers," *Wall Street Journal*, June 2, 2010; App Store sales estimate provided by Shaw Wu.

47.   **Meanwhile, Apple was soaring:** Interview with Jason Chen; Jesus Diaz, "How Apple Lost the iPhone 4," Gizmodo, April 19, 2010, http://gizmodo.com/5520438/how-apple-lost-the-next-iphone; "Police Seize Jason Chen's Computers," Gizmodo, April 26, 2010; Brian Lam, "Steve Jobs Was a Kind Man: My Regrets About Burning Him," Atlantic.com, October 6, 2011, http://www.theatlantic.com/technology/archive/2011/10/steve-jobs-was-a-kind-man-my-regrets-about-burning-him/246240/.

48.   **Hogan wanted ten thousand dollars:** In a Reddit thread published in June 2013 (http://www.reddit.com/r/IAmA/comments/1h2m81/i_leaked_the_iphone_4_ama/),

Hogan disputed Chen's account, claiming that Chen had agreed to pay $5,000 up front and another $3,000 upon confirmation by Apple that it was a real prototype. Hogan did not respond to the author's request to verify the authenticity of this account.

49. **Though Jobs did his best:** "Apple WWDC 2010 Keynote," Apple Keynotes podcast, June 7, 2010.

50. **Fans around the world:** Author's reporting around iPhone 4 launch; "Lab Tests: Why Consumer Reports Can't Recommend the iPhone 4," *Consumer Reports*, July 12, 2010; Isaacson, *Steve Jobs*, 521; Geoffrey A. Fowler, Ian Sherr, and Niraj Sheth, "A Defiant Steve Jobs Confronts 'Antennagate,'" *Wall Street Journal*, July 16, 2010; "Live Blogging Apple's Press Conference: Free Cases For All," WSJ.com, July 16, 2010, http://blogs.wsj.com/digits/2010/07/16/live-blogging-apples-press-conference/; "Song a Day #561: The iPhone Antenna Song," YouTube video posted by Jonathan Mann on July 15, 2010, http://www.youtube.com/watch?v=VKIcaejkpD4; Joshua Topolsky, "Live from Apple's iPhone 4 Press Conference," Engadget, July 16, 2010.

Chapter 4: Attila the Hun of Inventory

53. **That April:** "Appholes," *The Daily Show with Jon Stewart*, April 28, 2010, http://www.thedailyshow.com/watch/wed-april-28-2010/appholes.

54. **Apple's board recognized:** Isaacson, *Steve Jobs*, 518; author's interview with Walter Isaacson.

55. **"We have over:** The quote reflects the exact wording of the document, including Apple's inconsistent use of the uppercase *A* when spelling *apps*.

55. **Jobs still ruled over Apple:** Interviews with Greg Petsch, who had worked with Cook at Compaq; Lashinsky, "Tim Cook: The Genius Behind Steve," *Fortune*, August 24, 2011; Angie Lowry, "Thinking Different," *Auburn Magazine*, Winter 1999.

57. **Apple needed new blood:** Interviews with former and current Apple executives including Joe O'Sullivan, former operations team staff, and suppliers; Adam Lashinsky, *Inside Apple* (New York: Business Plus Hachette, 2012), 97; Lashinsky, "Tim Cook."

59. **The black laptop computer:** A more popular version of this tale places Sabih Khan's destination as China, but a person who was present at the meeting recalled that Khan had actually headed to Singapore.

59. **Later when "Wallstreet":** Interviews with former and current Apple executives including Joe O'Sullivan, former operations team staff, and suppliers; Lashinsky, *Inside Apple*, 97; Lashinsky, "Tim Cook."

62. **Cook's meticulous approach:** Interviews with former operations team staff, suppliers, and others with firsthand knowledge.

63. **Amid such tensions:** Interviews with suppliers and current and former Apple employees. Yukari Iwatani Kane, "For Apple Suppliers, Pressure to Win," *Wall Street Journal*, August 15, 2010; *Apple Inc. v. Paul Shin Devine et al.*, No. CV10-035633, Complaint, August 13, 2010, 14, 17. Paul Devine did not respond to an interview request.

64. **Apple's success brought pressure:** Interview with an insider with firsthand knowledge of Foxconn's suicide situation; David Barboza, "After Suicides, Scrutiny of China's Grim Factories," *New York Times*, June 6, 2010; David Barboza, "Electronics Maker Promises Review After Suicides," *New York Times*, May 26, 2010; Gordon G. Chang, "Suicides at Apple Supplier in China," *Forbes*, May 28, 2010; Long Kun and Li Wei, "Fu Shi Kang Yuan Gong Chang Qu Nei Si Wang, Jin Fang Chu Bu Pan Ding Xi Cu Si" [Foxconn Employee Died in Factory, Considered Preliminarily as Sudden Death by Police], trans. by Violet Tian, *Guangzhou Daily*, January 26, 2010.

Chapter 5: The Next Lily Pad

66.   **At dawn one brisk:** Interviews with multiple sources who attended the Top 100 and Bottom 100 meetings.

71.   **Jobs knew that his cancer:** Isaacson, *Steve Jobs*, 547–49.

71.   **One day in late 2010:** Interview with Gilbert Wong.

72.   **Wong may have been awed:** Isaacson, *Steve Jobs*, 547–49; "Apple Media Advisory," Apple press release, January 17, 2011; "Apple Reports First Quarter Results," Apple press release and earnings call, January 18, 2011.

74.   **That February, Jobs:** Interview with Tom Suiter.

77.   **Jobs kept the details of his health:** "Apple Special Event, March 2011," Apple Keynotes podcast, March 2, 2011; Isaacson, *Steve Jobs*, 552; author's interviews with Walt Mossberg, Walter Isaacson, Howard Stringer, and other people, whose names are withheld upon mutual agreement.

79.   **The next evening:** "Steve Jobs Presents to the Cupertino City Council (6/7/11)," YouTube video posted by Cupertino CityChannel on June 7, 2011; Isaacson, *Steve Jobs*, 535; LSA Associates Inc., "Apple Campus 2 Project Environmental Impact Report," June 2013; author's interview with Gilbert Wong.

80.   **By July, Job's cancer had spread:** Interview with sources with firsthand knowledge of the situation; Isaacson, *Steve Jobs*, 557; Josh Tyrangiel, "Tim Cook's Freshman Year: The Apple CEO Speaks," *BloombergBusinessweek*, December 6, 2012.

81.   **The date Jobs chose:** Isaacson, *Steve Jobs*, 557–58; "Letter from Steve Jobs," Apple press release, August 24, 2011.

82.   **Back home, Jobs:** Isaacson, *Steve Jobs*, 559; author's interviews with Walter Isaacson and Walt Mossberg.

84.   **Once Jobs announced his resignation:** Lashinsky, *Inside Apple*, 155; Charles Duhigg, "With Time Running Short, Steve Jobs Managed His Farewells," *New York Times*, October 7, 2011; Mona Simpson, "A Sister's Eulogy for Steve Jobs," *New York Times*, October 30, 2011.

84.   **The death certificate cited:** Interview with sources who communicated with executive team members or were inside Apple at the time; "Apple Media Advisory," Apple press release with Tim Cook's letter to employees about Steve Jobs's death, October 5, 2011, http://www.apple.com/pr/library/2011/10/05Apple-Media-Advisory.html; "Steve Jobs After the Resignation," TMZ, August 26, 2011, http://www.tmz.com/2011/08/26/steve-jobs-apple-photo-resignation-ceo-sick/; Steve Jobs's death certificate, County of Santa Clara, San Jose, California, October 5, 2011; Lauren Effron, "President Obama, Bill Gates, Mark Zuckerberg, Others React to Steve Jobs' Death," ABC News, October 5, 2011, http://abcnews.go.com/Technology/reaction-steve-jobs-death/story?id=14678187#.UX-3kmqsacwk.

Chapter 6: Ghost and Cipher

87.   **His ghost loomed everywhere:** Interviews with Walter Isaacson; Avie Tevanian, "Steve's Bachelor Party," Facebook (post), October 5, 2011, https://www.facebook.com/notes/avie-tevanian/steves-bachelor-party/10150860779725691; Brian Lam, "Steve Jobs Was Always Kind to Me (Or, Regrets of an Asshole)" The Wirecutter (blog), October 5, 2011, http://thewirecutter.com/2011/10/steve-jobs-was-always-kind-to-me-or-regrets-of-an-asshole/; author's interview with Walter Isaacson.

88.   **Even the rituals of remembrances:** Interviews with Tom Suiter and other

people who were present at the service; Nick Wingfield, "Memorial Service for Steve Jobs at Stanford," *New York Times*, Bits blog, October 14, 2011, http://bits.blogs.nytimes.com/2011/10/14/memorial-service-for-steve-jobs-planned-at-stanford/; Jessica E. Vascellaro and Ian Sherr, "Steve Jobs Memorial Held," *Wall Street Journal*, October 17, 2011

89. **The genius trap had been set:** Interviews with Mike Slade, Debbie Williams, John Underwood, current and former Apple executives and employees; Lowry, "Thinking Different"; Tim Cook, interview by Walt Mossberg and Kara Swisher, D11, All Things Digital, May 29, 2013, http://allthingsd.com/20130529/apples-tim-cook-the-full-d11-interview-video/; Lashinsky, "The Genius Behind Steve"; Donald and Geraldine Cook, interview by Debbie Williams, WKRG, January 16, 2009; Donna Riley-Lein, "Apple No. 2 Has Local Roots," *Independent*, December 25, 2008.

93. **For a man who craved invisibility:** Interviews with Robert Bulfin, Saeed Maghsoodloo, Charles Murphy, Barbara Davis, other Robertsdale High School former staff and students, Debbie Williams and Greg Petsch; Research by Pensacola Historical Society and the National Rural Electric Cooperative Association; Lowry, "Thinking Different"; Ray Garner, "Steve Jobs' World Man," *Business Alabama*, November 1999; Tim Cook, interview by Debbie Williams, WKRG, January 16, 2009; *Robala* (Robertsdale High School Yearbook), 1972–78; Ellen Williams, "Community News: Robertsdale," and other articles, *Independent*, January 1977–May 1978; Jeff Amy, "Tim Cook, Steve Jobs' Successor at Apple, Has Alabama Roots, Auburn Spirit," *Press-Register*, August 24, 2011; Paul Carroll, *Big Blues: The Unmaking of IBM* (New York: Crown, 1993), 59; *The Auburn Creed*; Auburn University, http://www.auburn.edu/main/auburn_creed.html; Miguel Helft, "The Understudy," *New York Times*, January 24, 2011; Tim Cook, interview by Duke University, April 2013, https://www.youtube.com/playlist?list=PLwEToxwSycW1uqGG-iYZOERU0WBTKIAMt.

97. **Decades later:** Interviews with Joe O'Sullivan and multiple former Apple executives and employees; Tim Cook, "Auburn University 2010 Commencement Speech," May 14, 2010, YouTube, http://www.youtube.com/watch?v=xEAXuHvzjao; Adam Lashinsky, "How Tim Cook Is Changing Apple," *Fortune*, May 24, 2012; Lowry, "Thinking Different"; Tim Cook, interview by Duke University, April 2013, https://www.youtube.com/playlist?list=PLwEToxwSycW1uqGG-iYZOERU0WBTKIAMt; "The Power List 2013," *Out*, April 10, 2013. Underwear comment was based on a witness's firsthand observation.

104. **Jobs's departure presented a crisis:** Interviews with current and former Apple employees and executives; "Apple Media Advisory," Apple press release, October 5, 2011; Lashinsky, "How Tim Cook is Changing Apple"; Lowry, "Thinking Different."

Chapter 7: Joy City

108. **Sun Danyong had been groomed:** "Diu Shi iPhone Bei Diao Cha, Fu Shi Kang Yuan Gong Tiao Lou Zi Sha" [iPhone Lost, Foxconn Staff Being Investigated Committed Suicide], trans. by Violet Tian, *Southern Metropolis Daily*, July 21, 2009; "Sun Danyong Shi Jian Jing Dong Guo Taiming—Ben Bao Du Jia Na Dao Dang Ri Quan Cheng Jian Kong Lu Xiang, She Shi Ke Zhang Tu Lu Xin Sheng" [Terry Gou Disturbed by Sun Danyong's Suicide—Exclusive Reportage on CCTV Record on the Day of Suicide and Confession of the Involved Head of Department], trans. by Tian, *Southern Metropolis Daily*, July 22, 2009; Xie Peng and Zhai Qiaohong, "Duo Ming Shou Ji: Fu Shi Kang vs. Da Xue Sheng Sun Danyong Qi Ye Wen Hua vs. Yuan Gong Ren Xing" [A Lethal Phone: Foxconn vs. College Graduate, Sun Danyong and Enterprise Culture vs. Employee's Humanity], trans. by Violet Tian, *Southern Weekend*, July 30, 2009; Evan Osnos, "Death at an Apple Manufacturer in China," *New Yorker*, July 23, 2009; Evan Osnos, "More on the iPhone Suicide," *New Yorker*, July 24, 2009.

109.   **It's unclear where he went:** Chat record between Sun and his friends on QQ, trans. by Tian; http://www.fenfenyu.com/Memorial_Static/2901/Article/4.html; "Fu Shi Kang Si Wang Yuan Gong Tong Xue: 'Tie Zi Bu Shi Wo Fa De'" [Classmate of Dead Foxconn Staff: "I Didn't Post It Online"], trans. by Tian, *Guangzhou Daily*, July 22, 2009.

110.   **The twenty-two-year-old:** Many reports said Sun was twenty-five years old, but his friend's blog claimed his birthday was on September 1, 1986, which made him twenty-two. This is consistent with his 2004 college admission year.

111.   **When Sun's suicide became public:** Interviews with Jason Dean, Louis Woo, local reporter, whose name is undisclosed to protect from persecution, and a person with firsthand knowledge of Foxconn's media handling of the suicide; "The Other Side of Apple," Friends of Nature, IPE, Green Beagle, January 20, 2011; Charles Duhigg and Keith Bradsher, "How the U.S. Lost Out on iPhone Work," *New York Times*, January 21, 2012; Jason Dean, "The Forbidden City of Terry Gou," *Wall Street Journal*, August 11, 2007; Gou, Tai-Ming, "Jie Ma Guo Taiming Yu Lu—Chao Yue Zi Wo De Yu Yan" [Deciphering Gou's Quotations—Prophesies that Overcome Oneself], (Taiwan: TianXiaWenHua, 2008); "Sun Danyong Shi Jian Jing Dong Guo Taiming—Ben Bao Du Jia Na Dao Dang Ri Quan Cheng Jian Kong Lu Xiang, She Shi Ke Zhang Tu Lu Xin Sheng" [Terry Gou Disturbed by Sun Danyong's Suicide—Exclusive Reportage on CCTV Record on the Day of Suicide and Confession of the Involved Head of Department], *Southern Metropolis Daily*, Online Issue SA29, July 22, 2009.

115.   **For a long time, the world:** Interviews with Pennee Saingarm about Shenzhen background as well as former Apple employees and executives with firsthand knowledge of the situation; "The Stark Reality of Chinese Factories," *Daily Mail*, August 18, 2006; "Report on iPod Manufacturing," Apple press release, August 17, 2006; Arnold Kim, "iPhone Factory Worker Photos Found on New iPhone," MacRumors, August, 20, 2008.

119.   **Around the same time:** Interview with Ma Jun; "The IT Industry Has a Critical Duty to Prevent Heavy Metal Pollution," Friends of Nature, Institute of Public and Environmental Affairs, Green Beagle, April 24, 2010, http://www.ipe.org.cn/Upload/Report-IT-Phase-One-EN.pdf; "The Other Side of Apple," Friends of Nature, IPE, Green Beagle, January 20, 2011.

121.   **On May 20, 2011:** Interviews with Ma Jun, Linda Greer, former Apple employees familiar with the situation; "The Other Side of Apple II: Pollution Spreads Through Apple's Supply Chain," YouTube video, posted by "somedayfire," January 10, 2012, http://www.youtube.com/watch?v=rpFz9VAX8zM; James T. Areddy and Yukari Iwatani Kane, "Explosion Kills 3 at Foxconn Plant," *Wall Street Journal*, May 21, 2011; "61 Workers Injured in Explosion at Shanghai Apple Supplier," China Labor Watch press release, December 19, 2011; Clare Jim and Argin Chang, "Apple Supplier Pegatron Hit by China Plant Blast," Reuters, December 19, 2011.

124.   **Finally the Western media:** "Mr. Daisey and the Apple Factory," *This American Life*, WBEZ, January 6, 2012; Ira Glass, "Retracting Mr. Daisey and the Apple Factory," *This American Life* (blog), WBEZ, March 16, 2012; Charles Duhigg and David Barboza, "In China, Human Costs Are Built Into an iPad," *New York Times*, January 25, 2012.

126.   **The exposé, a direct hit:** Interviews with people who were at the *New York Times*–Apple editorial meeting as well as current and former Apple executives with firsthand knowledge of other details discussed in this section; the *New York Times* did not respond to a request for comment on the editorial meeting; Killian Bell, "Tim Cook Visits Beijing to Meet Chinese Officials, Pose With Fans," *Cult of Mac*, March 26, 2012, http://www.cultofmac.com/156220/tim-cook-visits-beijing-to-meet-chinese-officials-pose-

with-fans/; Li Qiling, "Ku Ke Lai Hua Bei Hou De Xuan Ji" [Decoding Cook's Trip to China], trans. by Violet Tian, *Beijing News*, April 11, 2012, http://epaper.bjnews.com.cn/html/2012-04/11/content_330210.htm?div=-1.

Chapter 8: Into the Fire

129. **On an unseasonably warm day:** "Olympic Torch/Knighthood/Royal Academy Party," YouTube video, posted by "MyDigitalRealm," May 23, 2012, http://www.youtube.com/watch?v=kKMfnu4GFgY; "Apple designer Sir Jonathan Ive Knighted at Palace," BBC, May 30, 2012. Background on knighthood ceremony from Sir Howard Stringer.

130. **Ive's knighthood was one:** Interviews with former Apple executives familiar with both Jobs and Cook; Tim Cook, "A Celebration of Steve's Life," Apple video, October 19, 2011.

132. **To realize the magnitude of Cook's challenge:** Interviews with former Apple executives and employees; Graeme Wearden, "Networking Unnecessary for Rising Star Browett," *Guardian*, June 7, 2007; Philip Schiller, Twitter feed, https://twitter.com/pschiller.

134. **Scott Forstall's nickname:** Interviews with former Apple executives and employees, Forstall friends; Lashinsky, *Inside Apple*, 105–06; Adam Santariano, Peter Burrows, and Brad Stone, "Scott Forstall, the Sorcerer's Apprentice at Apple," *Bloomberg-Businessweek*, October 12, 2011.

136. **The world disagreed:** Interviews with Isaacson; Ive's former teachers Netta Cartwright and David Whiting; former professors, friends, pre-Apple colleagues including Barrie Weaver, Clive Grinyer, Peter Phillips, Martin Darbyshire; and former Apple colleagues including Robert Brunner; background on Newcastle provided by John Elliott and Steven Kyffin; Jonathan Ive, foreword to *Dieter Rams: As Little Design as Possible* (London: Phaidon, 2011); Paul Kunkel, *AppleDesign: The Work of the Apple Industrial Design Group*; Rob Waugh, "How Did a British Polytechnic Graduate Become the Design Genius Behind 200 Billion Pound Apple?," *Mail Online*, March 19, 2011; Isaacson, *Steve Jobs*, 342, 350.

140. **Hanging on to this team:** Interview with Apple colleague; Jonathan Ive, interview by James Naughtie, *Today*, BBC, May 24, 2012.

Chapter 9: Looks Like Rain

142. **When the iPhone 4S went on sale:** Interview with John Keitz; "Siri Won't Call My Dad," forum discussion by "jkeitz," "GingerSnapsBack" and others, iMore, July 6, 2012, http://forums.imore.com/iphone-4s/236834-siri-wont-call-my-dad.html.

143. **Expectations for the iPhone 4S:** Robert Cyran, "Why Apple Might Just Be the First $1 Trillion Company," Reuters, August 9, 2011. "iPhone 4S Launch: As It Happened," *Guardian* (blog), October 4, 2011, http://www.theguardian.com/technology/2011/oct/04/iphone-5-launch-live-coverage; "Live Blog: Apple Event—iPhone Announcement," *Wall Street Journal* (blog), October 4, 2011; "Apple Special Event October 2011," Apple video, October 4, 2011, http://www.apple.com/apple-events/october-2011/.

146. **At first, Siri was a sensation:** Interview with Nicky Kelly, Geoffrey Fowler, Karen Jacobsen, Adam Cheyer, and another person with firsthand knowledge; Bianca Bosker, "Siri Rising: The Inside Story of Siri's Origins—And Why She Could Overshadow the iPhone," *Huffington Post*, January 22, 2013; Matt Warman, "The Voice Behind Siri Breaks His Silence," *Telegraph*, November 10, 2011; Jessica Ravitz, "I'm the Original Voice of Siri," *CNN*, October 8, 2013.

148.   **The British Siri's voice belonged to Jon Briggs:** Briggs did not respond to a request to verify the details of his interaction with Apple

148.   **Despite Siri's futuristic promise:** Interview with Frank Meehan, Adam Cheyer, people with firsthand knowledge; Megan O'Neill, "Watch: Siri Doesn't Understand Foreign Accents," *SocialTimes*, October 17, 2011; Leslie Anne Harrison, "I Need a Southern Siri," GulfCoastNewsToday.com, March 19, 2013; Mark Gongloff, "Apple: Siri Goes AWOL, Stock Dips," *Wall Street Journal* (blog), October 17, 2011; Bryan Fitzgerald, "Woz Gallops in to a Horse's Rescue," *Times Union*, June 13, 2012; Bosker, "Siri Rising"; Gene Munster, "Apple's Siri a Quick Study, But Google at Head of the Class," *Piper Jaffray*, June 29, 2012; Erick Schonfeld, "Silicon Valley Buzz: Apple Paid More than $200 Million for Siri to Get into Mobile Search," TechCrunch, April 28, 2010.

153.   **Determined to make the new iPhone a hit:** "Martin Scorsese iPhone 4S Siri Commercial HD," YouTube video, posted by "Mareese Smith," August 13, 2012, http://www.youtube.com/watch?v=2t8O_92G7OU; "Zooey Deschanel iPhone 4S Siri Commercial," YouTube video, posted by "smallfries2," January 3, 2013, http://www.youtube.com/watch?v=fbEjCvdGaZU; "The Phone Store," Vimeo video, posted by Neal Desai, http://vimeo.com/53747757; Mitch Albom, "A Siri-ous Disconnect," *Reader's Digest*; Fitzgerald, "Woz Gallops in to a Horse's Rescue."

156.   **At this stage in Apple's evolution:** Interviews with Avie Tevanian, Gene Munster, people with knowledge of the situation including Apple executives; "Apple CEO Tim Cook at Goldman Sachs," transcript by Philip Elmer-DeWitt, *Fortune* (blog), February 15, 2012; George Colony, "Apple=Sony," Forrester (blog), April 25, 2012.

Chapter 10: Thermonuclear

160.   **The two industrial superpowers stood:** Interview with a former Apple executive with firsthand knowledge; *Apple Inc. v. Samsung Electronics*, No. 11-CV-01846, transcript, August 10, 2012; Ina Fried, "Apple Offered to License Its Patents to Samsung for $30 per Smartphone, $40 per Tablet," *All Things Digital*, August 10, 2012.

161.   **As a newly crowned titan:** Interviews with people with firsthand knowledge of the situation, including former Apple executives, Justice Department official, and publishing industry executives; John Markoff, "The Passion of Steve Jobs," *New York Times* (blog), January 15, 2008; Isaacson, *Steve Jobs*, 503–04; *United States v. Apple Inc.*, No. 12-cv-02826, Opinion & Order, July 10, 2013.

165.   **As Apple plotted to remake:** Interview with Florian Mueller and other patent experts; Ian Edmonson and Yukari Iwatani Kane, "Nokia Accuses Apple of Patent Infringement," *Wall Street Journal*, October 23, 2009; size estimate of Nokia's patent portfolio size provided by Florian Mueller.

167.   **The philosophy behind Android:** Interview with former Apple executives familiar with the situation; Fred Vogelstein, *Dogfight: How Apple and Google Went to War and Started a Revolution* (New York: Sarah Crichton Books, 2013); Jessica E. Vascellaro and Yukari Iwatani Kane, "Apple, Google Rivalry Heats Up," *Wall Street Journal*, December 10, 2009; John C. Abell, "Google's 'Don't Be Evil' Mantra Is 'Bullshit,' Adobe Is Lazy: Apple's Steve Jobs," *Wired* (blog), January 30, 2010; "2004 Founders' IPO Letter," Google, August 18, 2004, http://investor.google.com/corporate/2004/ipo-founders-letter.html; Arnold Kim, "Steve Jobs at Apple Town Hall Meeting on Google, Adobe, Next iPhone, 2010 Macs and More," MacRumors, January 30, 2010; Isaacson, *Steve Jobs*, 512–13.

169.   **"They want to kill the iPhone:"** This quote was also disputed in *Daring Fireball*, which reported that Jobs actually said, *"teams at* Google want to kill us." The author

elected to go with the version in *Wired* because the publication had the fullest account of the entire meeting.

169.   "'Don't be evil' is a load of crap": *Wired* originally quoted Jobs as saying, "This don't be evil mantra: it's bullshit," but the publication later posted an update saying that another source corrected the quote. The author used the corrected version, which was also reported by *Daring Fireball*.

171.   The lawsuit had clearly come: Interview with former Apple executive familiar with the situation.

172.   On the evening after: Interviews with Ben Fullerton, J. P. Stallard, Clive Grinyer, a former Apple executive and a Samsung engineer whose names were withheld upon mutual agreement; Yukari Iwatani Kane and Ian Sherr, "Apple: Samsung Copied Design," *Wall Street Journal*, April 19, 2011; Evan Ramstad, "Samsung Sues Apple on Patents," *Wall Street Journal*, April 22, 2011; "Android Rises, Symbian 3 and Windows Phone 7 Launch as Smartphone Shipments Increase 87.2% Year Over Year, According to IDC," IDC press release, February 7, 2011; "Apple Management Discusses Q2 2011 Results," Transcript, Seeking Alpha, April 20, 211.

175.   Like Apple, Samsung: Reporting by Yoo-Chul Kim; Sea-Jin Chang, *Sony vs. Samsung: The Inside Story of the Electronics Giants' Battle for Global Supremacy* (Singapore: John Wiley & Sons (Asia), 2008); Tony Mitchell, *Samsung Electronics and the Struggle for Leadership of the Electronics Industry* (Singapore: John Wiley & Sons (Asia), 2010); Frank Rose, "Seoul Machine," *Wired*, May 13, 2005, http://www.wired.com/wired/archive/13.05/samsung.html?pg=1&topic=samsung&topic_set; Boon-Young Lee and Seung-Joo Lee, "Case Study of Samsung's Mobile Phone Business," KDI School of Public Policy and Management, 2004.

177.   Between HTC, Motorola: Interviews with people with firsthand knowledge of the situation; Larry Dignan, "Nokia Likely Netted $600 Million Plus in Apple Settlement," ZDNet, June 14, 2011; Christopher Lawton and Dominic Chopping, "Nokia, Apple Make Up," *Wall Street Journal*, June 15, 2011; Florian Mueller, "Is Apple Winning or Losing the Patent Game?," Foss Patents (blog), May 19, 2011; estimate of lawsuits between Apple and Samsung by Florian Mueller; "Yeoyoumanman Samsung'Apple gua so-song, sonhaebol geot upda,'" [Samsung: "Nothing to Lose from Lawsuit Against Apple"], trans. by MinSun Lee, *Hankyoreh*, August 31, 2011; Kelly Olsen, "Samsung to Step Up Apple Patent War," Associated Press, September 23, 2011; "Lee Jae-young, Tim Cook heoidong . . . Samsung, Apple daetahyup?" [Jae Lee and Tim Cook Meet . . . Samsung and Apple's Great Compromise], trans. by MinSun Lee, *Seoul Shinmun*, October 17, 2011; Kim Yoo-chul, "It's All About War, not Battles, Samsung Says," *Korea Times*, October 17, 2011; Eric Slivka, "Steve Jobs Exhibit on Display at U.S. Patent Office Museum," MacRumors, November 23, 2011; *Apple Inc. v. Samsung Electronics*, No. 11-CV-01846, transcript, October 13, 2011; *Apple Inc. v. Samsung Electronics*, No. 11-CV-01846, Order Denying Motion for Preliminary Injunction, December 2, 2011. All quotes from this trial and others here and in the rest of the book are based on official transcripts.

183.   Choi Gee-sung: Also known as G.S. Choi or Geesung Choi in English.

Chapter 11: The Innovator's Dilemma

185.   Three thousand miles away: Interviews with Clayton Christensen, Gautam Mukunda, Hal Gregersen, and Apple insiders; Clayton Christensen, *The Innovator's Dilemma* (New York: Harper Business, 2011); Jena McGregor, "Clayton Christensen's Innovation Brain," *BusinessWeek*, June 15, 2007; Larissa Macfarquhar, "When Giants Fail," *New Yorker*, May 14, 2012; Clayton Christensen, interview by Horace Dediu,

Asymco, May 2, 2012, http://www.asymco.com/2012/05/02/5by5-the-critical-path-36-an-interview-with-clayton-christensen/; "Android and iOS Surge to New Smartphone OS Record in Second Quarter, According to IDC," IDC press release, August 8, 2012; Tim Cook, interview by Walt Mossberg and Kara Swisher, D10, All Things Digital, May 29, 2012.

Chapter 12: Boundless Oceans, Vast Skies

196. **As dusk descended on:** Interview with Ai Qi and other factory workers, Beyond, "Boundless Oceans, Vast Skies," trans. by Huey's Stuff (blog), November 1, 2011, http://hueyly.com/wordpress/beyond-hoi-foot-tin-hon.

198. **Students and Scholars Against Corporate Misbehaviour:** SACOM is often criticized by corporations for its extreme views. Based on an hour interview with its leader Debby Sze Wan Chan, I determined that their sample pool of interviewees was too small and biased against Foxconn to derive a fair overall picture of the conditions at the factory. SACOM volunteers identify themselves as labor activists, which means most workers avoid them unless they are particularly disgruntled. However, I have found the group's anecdotes and facts to be consistent with other accounts and believe their accounts to be honest and truthful.

200. **For a long time, Ai had:** Interview with Ai Qi. Exchange rates may vary because they are calculated based on the rate in the timeframe mentioned.

203. **Foxconn did what it could:** Interview with Ai Qi; Jason Dean, "The Forbidden City of Terry Gou," *Wall Street Journal*, August 11, 2007; "Global 500," *Fortune*, July 25, 2011; Frederik Balfour and Tim Culpan, "The Man Who Makes Your iPhone," *Bloomberg-Businessweek*, September 9, 2010; *Foxconn People* newsletter, trans. by Violet Tian, June 7, 2012 issue; "Fair Labor Association Secures Commitment to Limit Workers' Hours, Protect Pay at Apple's Largest Supplier," Fair Labor Association press release, March 29, 2012.

205. **Around that time:** Interview with Debby Sze Wan Chan, Geoffrey Crothall; "Apple's CEO Discusses Q4 2011 Results—Earnings Transcript," Seeking Alpha, October 18, 2011; "2012 National Economic and Social Development Report," trans. by Violet Tian, National Bureau of Statistics of China, February 22, 2013, http://www.stats.gov.cn/tjgb/ndtjgb/qgndtjgb/t20130221_402874525.htm; "Income of Urban and Rural Residents in 2011," National Bureau of Statistics of China press release, January 30, 2012; Lu Mai, "Poverty Eradication in China: A New Phase," China Development Research Foundation, February 10, 2011, http://www.un.org/esa/socdev/csocd/2011/Lu.pdf; Foxconn salary data by Students and Scholars Against Corporate Misbehaviour; "Fair Labor Association Secures Commitment to Limit Workers' Hours, Protect Pay at Apple's Largest Supplier," Fair Labor Association press release, March 29, 2012.

206. **The average per capita income:** Exchange rate is as of December 30, 2011.

206. **What saved Ai from:** Interview with Ai Qi; Beyond, "Boundless Oceans, Vast Skies"; Pun Ngai, *Made in China* (Durham: Duke University Press, 2005), 189–96; hardware engineer salary, Indeed, based on May 2013 data, http://www.indeed.com/salary/Hardware-Engineer.html.

Chapter 13: Fight Club

210. **The trial between Apple and Samsung:** Interviews with Florian Mueller, Lea Shaver, other patent experts; Dan Levine, "Apple and Samsung Execs in Talks on Patent Lawsuits," Reuters, June 17, 2011; "Apple and Samsung Garner 50% of Global Smartphone Market and 90% of Its Profits," ABI Research press release, June 15, 2012.

212. **The public war raged on television:** Interviews with sources with firsthand knowledge; "Next Big Thing," Samsung 2012 Super Bowl ad, posted by *Adweek*, January 26, 2012, http://www.adweek.com/video/advertising-branding/samsung-next-big-thing-2012-super-bowl-teaser-137740; Sean Hollister, "Samsung Says Apple Lawsuit Didn't Affect Its Smartphone Design," *The Verge*, May 22, 2012; Miyoung Kim, "Insight: Samsung: 'Fast Executioner' Seeks Killer Design," Reuters, May 23, 2012; Ina Fried, "Samsung, Apple Even at Odds Over Where They Will Sit at Trial," *All Things D*, July 26, 2012, http://allthingsd.com/20120726/samsung-apple-even-at-odds-over-where-they-will-sit-at-trial/; *Apple Inc. v. Samsung Electronics Co.*, No. 11-cv-01846, (Disputed) Joint Proposed Jury Instructions, July 24, 2012; Stephen Lawson, "Judge Again Orders Apple, Samsung to Streamline Claims in iPad Patent Case," IDG News Service, May 2, 2012; Matt Macari, "Apple Gives Samsung Some Work-Around Options for its iPhone and iPad Design Patents," *The Verge*, December 2, 2011.

215. **On July 30, 2012:** Author's personal observations from attending the trial; interviews with sources with firsthand knowledge of the trial, including Reuters reporter Poornima Gupta; *Apple Inc. v. Samsung Electronics Co.*, No. 11-cv-01846, transcript, July 30, 2012; Shannon Green, "Winning: Charles Verhoeven," *National Law Journal*, June 21, 2010.

218. **Opening arguments began:** Interviews with people with firsthand knowledge of the trial as well as Manuel Ilagan; *Apple Inc. v. Samsung Electronics Co.*, No. 11-cv-01846, transcripts, July 31, August 3, August 6, August 14, 2012; Rebecca English, "Elegant Kate goes for Dove Grey and a New Side Chignon Up-do as She Steps Out Solo for Cocktail Reception," *Daily Mail*, July 30, 2012; "Samsung Calls BS on Apple's Charges of Copying," Conan, August 7, 2012, http://teamcoco.com/video/samsung-denies-apples-charges-copying.

224. **Samsung was cornered:** *Apple Inc. v. Samsung Electronics Co.*, No. 11-cv-01846, transcripts, July 31, August 3, August 15, August 21, 2012.

227. **In closing statements:** *Apple Inc. v. Samsung Electronics Co.*, No. 11-cv-01846, transcript, August 21, 2012.

228. **In complex cases like this one:** Interviews with Martyn Williams, Manuel Ilagan, other people with firsthand knowledge of the trial; *Apple Inc. v. Samsung Electronics Co.*, No. 11-cv-01846, transcript, August 24, 2012; Howard Mintz, "Jury Foreman in *Apple v. Samsung*: Verdict a Message That Copying Is a Big Risk," *San Jose Mercury News*, August 25, 2012; Jessica E. Vascellaro, "Inside the Apple-Samsung Jury Room," *Wall Street Journal*, August 27, 2012; Mark Gurman, "Tim Cook Tells Apple Employees That Today's Victory 'Is About Values,'" 9to5Mac, August 24, 2012; Connie Guglielmo, "Apple Wins Over Jury in Samsung Patent Dispute, Awarded $1.05 Billion in Damages (Live Blog)," *Forbes*, August 24, 2012.

230. **In Munich, the patent expert:** Interview and email correspondence with Florian Mueller; Florian Mueller, "Apple's Billion-Dollar Win over Samsung Is a Huge Breakthrough—but It's Not Thermonuclear," Foss Patents (blog), August 24, 2012; Florian Mueller, Twitter update, August 24, 2012, https://twitter.com/FOSSpatents/status/239181522761027584.

Chapter 14: Typhoon

232. **Days after the verdict:** Interview with Samsung executive; reporting by Kim Yoo-chul and Kim Hyung-eun; Kim Yoo-chul, "Samsung Scrambles to Recover after Uppercut," *Korea Times*, August 26, 2012; "Choi Ji-sung, guteun pyojeongeuro chulgeun . . . Samsung gingeup daechaek heoieui" [Choi Ji-sung, coming to work with stiff look . . . Samsung to hold emergency meeting], trans. by MinSun Lee, *Asia Gyeongje*,

August 26, 2012; "Lee Kun-hee heoijang 'Apple teukhu sosong jal hara,' " [Chairman Lee Kun-hee "Do a Good Job with Patent Lawsuit Against Apple"], trans.by MinSun Lee, *Newsis News Agency*, August 29, 2012; Seongjin Cha and Rose Kim, "Typhoon Bolaven Capsizes Fishing Boats, Lashes Korea," *BloombergBusinessweek*, August 28, 2012; "Bolaven Leaves at Least 16 Dead as It Plows Past Korean Peninsula," CNN, August 29, 2012; Kim, Chang-woo, "Chuijae Ilgi: California basimwoneui 'miguksik jungeui,' " [Reporter's Note: California Jury's "American Style Justice"], trans. by MinSun Lee, *Joongang Ilbo*, August 28, 2012.

233.   **Apple's lawyers had no illusions:** Interviews with sources with firsthand knowledge of post-trial details; Charles Duhigg and Steve Lohr, "The Patent, Used as a Sword," *New York Times*, October 7, 2012; Steve Lohr, "Why Apple Might be Better Off Losing Its Patent Lawsuit," *New York Times*, Bits blog, August 21, 2012, http://bits.blogs. nytimes.com/2012/08/21/why-apple-might-be-better-off-losing-its-patent-lawsuit/.

235.   **Apple was plotting:** Interview with a source familiar with the situation; Mark Gurman, "Apple Acquired Mind-Blowing 3D Mapping Company C3 Technologies Looking to Take iOS Maps to the Next Level," 9to5Mac, October 29, 2011, http://9to5mac. com/2011/10/29/apple-acquired-mind-blowing-3d-mapping-company-c3-technologies-looking-to-take-ios-maps-to-the-next-level/; "Apple Special Event," Apple video, June 11, 2012, http://www.apple.com/apple-events/june-2012/; Jessica E. Vascellaro and Amir Efrati, "Apple and Google Expand Their Battle to Mobile Maps," *Wall Street Journal*, June 4, 2012.

237.   **Apple's growing defensiveness:** Interviews with reporters who were present at the event including Ian Sherr and Nobuyuki Okada; "Apple Special Event," Apple video, September 12, 2012, http://www.apple.com/apple-events/september-2012/; Dakster Sullivan, "iPhone 5 Review—What's Up and What's Not," *Wired*, October 15, 2012, http:// www.wired.com/geekmom/2012/10/iphone-5-review/; Juro Osawa, "Apple Cuts Orders for iPhone Parts," *Wall Street Journal*, January 13, 2013.

240.   **Apple could not afford:** Interview with Markus Thielking, people familiar with the situation; Mike Butcher, "Welcome to Apple's iOS6 Map—Where Berlin Is Now Called 'Schoeneiche,' " TechCrunch, September 20, 2012, http://techcrunch.com/2012/09/20/ welcome-to-apples-ios6-map-where-berlin-is-now-called-schoeneiche/; Eric Mack, "Apple Maps Fiasco Makes *Mad*'s '20 Dumbest' List," CNET, December 6, 2012; Josh Lowensohn, "Developers: We Warned Apple About iOS Maps Quality," CNET, October 9, 2012; "Letter from Bob Mansfield," Apple press release, July 13, 2012, http://www.apple. com/environment/letter-to-customers/; "A Letter to Our Customers Regarding Maps," Apple press release, September 28, 2012, http://www.apple.com/letter-from-tim-cook-on-maps/; Isaacson, *Steve Jobs*, 522; Jessica E. Lessin, "An Apple Exit Over Maps," *Wall Street Journal*, October 29, 2012; Josh Tyrangiel, "Tim Cook's Freshman Year: The Apple CEO Speaks," *BloombergBusinessweek*, December 6, 2012; background on user interface design by Ben Fullerton.

240.   **street matching:** Also known as map matching.

244.   **John Browett was another:** Interview with Gary Allen; Gary Allen, "Browett Leaves, Third Party Shelves Get Make Over," ifo Apple Store (blog), November 4, 2012, http://www.ifoapplestore.com/2012/11/04/browett-leaves-third-party-shelves-get-make-over/; Ian Sherr, "Apple Retail Chief Admits Staffing Mistake," *Wall Street Journal*, August 16, 2012; Ben Lovejoy, "John Browett Reflects on Lessons from Brief Stint as Apple Retail Chief," MacRumors, March 15, 3013.

246.   **One of them was a Harvard professor:** Interviews with Gautam Mukunda, former and current Apple executives; Lashinsky, *Inside Apple*, 161; Gautam Mukunda, *In-*

dispensable: When Leaders Really Matter (Boston: Harvard Business Review Press, 2012), 1–19.

Chapter 15: Revolt

250. **Two days after the iPhone 5:** Interviews with Foxconn workers, including Wang Sheng and others familiar with the situation; "Fu Shi Kang Taiyuan Yuan Gong Chong Tu Bei Hou: Yan Jun De Guan Li Tiao Zhan" [Behind Foxconn's Taiyuan Strikes: Severe Management Challenges], trans. by Violet Tian, *China Business News*, September 25, 2012; "Foxconn Workers Labor Under Guard After Riot Shuts Plant," Bloomberg, September 26, 2012; "3000 to 4000 Workers Strike at Foxconn's China Factory," China Labor Watch press release, October 5, 2012; "Foxconn Denies China iPhone Plant Hit by Strike," Reuters, October 6, 2012; "Fu Shi Kang Gao Ceng Duan Xin Zhao Yuan Gong Fu Gong, Yuan Gong Mei Tian Zhan 11 Ge Xiao Shi" [Senior Foxconn Staff Texted Workers to Get Back to Work, Workers Kept Standing for as Long as 11 Hours Daily], trans. by Violet Tian, *Securities Daily*, October 8, 2012; Wang Sheng/yefudao's Sina Weibo page, http://weibo.com/u/2175973444?topnav=1&wvr=5&topsug=1; Liu Yong, "Fu Shi Kang Zhengzhou Zhi Luan: Su Fang Guan Li Yin Fa Chong Tu" [The Chaos in Foxconn Zhengzhou: Extensive Management Gives Rise to Conflicts], *China Business Journal*, October 13, 2012.

253. **The workers' riot and the strike:** Wang Yu Wen, "Guo Tai Ming Wangchao Quanbuxi," [The Empire of Terry Gou, A Full Dissection], trans. by Kristen Choy, *Shangye Zhoukan* [Taiwan Business Weekly], January 16, 2013; "Fortune 500," *Fortune*, http://money.cnn.com/magazines/fortune/fortune500/2012/full_list/.

255. **Over the years:** Interviews with Greg Petsch and others with firsthand knowledge of Foxconn's inner workings, including former Foxconn/Hon Hai employees, partners, and consultants; journalist-consultant Tomorhiro Otsuki also provided background on Foxconn. Frederik Balfour and Tim Culpan, "The Man Who Makes Your iPhone," *BloombergBusinessweek*, September 9, 2010; Naoyoshi Goto and Jun Morikawa, *Apple Teikoku no Shotai* [The Apple Empire's True Self], (Tokyo: Bungeishunjyu, 2013); Ken Takeuchi, *Sekai De Shobu Suru Shigotojutsu* [Globally Competitive Work Strategies] (Japan: Gentosha, 2012), 81–82; Wang Yu Wen, "Guo Tai Ming Wangchao Quanbuxi" [The Empire of Terry Gou: A Full Dissection], trans. by Kristen Choy, *Shangye Zhoukan* [Taiwan Business Weekly], January 16, 2013; Lin Yang, "Foxconn Tries to Move Past the iPhone," *New York Times*, May 6, 2013; Lorraine Luk, "As Apple Feels Bite, Hon Hai Looks to Diversify," *Wall Street Journal*, May 27, 2013.

260. **Apple recognized the danger:** Interviews with people with firsthand knowledge of Apple's partnership and others, including former Foxconn/Hon Hai employees, partners, and consultants; Monica Chen and Joseph Tsai, "Pegatron to See Strong 1Q13 Due to iPad mini Orders," *DigiTimes*, December 17, 2012, http://www.digitimes.com/news/a20121217PD215.html; Goto and Morikawa, *Apple Teikoku no Shotai*, 26–28; Lan Guanming, "Bu Chi Ping Guo De Li You" [The Reasons Why We Don't Eat Apple], trans. by Violet Tian, *CTimes*, April 5, 2012.

264. **In the last weeks of 2012:** Spencer E. Ante and Will Connors, "In the Smartphone Race, Money Talks for Samsung," *Wall Street Journal*, March 12, 2013; Juro Osawa, "Apple Cuts Orders for iPhone Parts," *Wall Street Journal*, January 14, 2013; Goto and Morikawa, *Apple Teikoku no Shotai*, 19; "Monozukuri Nihon ni hirogaru Apple shihai" [The Broadening Domination of Apple in Manufacturing Japan], *Diamond Weekly*, October 6, 2012; Lorraine Luk, "Hon Hai Plots Next Move," *Wall Street Journal* (blog), June 27, 2013, http://blogs.wsj.com/digits/2013/06/27/hon-hai-plots-next-move/; reporting about Sharp provided by Naoyoshi Goto.

Chapter 16: Velvin

266. **The jury foreman would not:** Dan Levine, "Jury Didn't Want to Let Samsung Off Easy in Apple Trial: Foreman," Reuters, August 25, 2012; Howard Mintz, "Jury Foreman in *Apple v. Samsung*: Verdict a Message That Copying Is a Big Risk," *San Jose Mercury News*, August 25, 2012; Bryan Bishop, "*Apple v. Samsung* Jury Foreman: Only the 'Court of Popular Opinion' Can Change the Patent System," *Verge*, August 31, 2012; Mario Aguilar, "Ask Apple vs Samsung Jury Foreman Velvin Hogan Whatever You Want," Gizmodo, September 4, 2012.

268. **Venomous as the free-for-all:** Interviews with people familiar with the situation and Florian Mueller; Velvin Hogan, interviewed by Emily Chang, Bloomberg Television, August 27, 2012; Joel Rosenblatt, "Samsung Claims Jury Foreman Misconduct Tainted Apple Case," Bloomberg, October 3, 2012; *Apple Inc. v. Samsung Electronics Co.*, No. 11-cv-01846, transcript, December 6, 2012; *Apple Inc. v. Samsung Electronics Co.*, No. 11-cv-01846, Order Denying Motion for Permanent Injunction, December 17, 2012; *Apple Inc. v. Samsung Electronics Co.*, No. 11-cv-01846, Order Re: Juror Misconduct, December 17, 2012; *In the Matter of Certain Electronic Devices Including Wireless Communication Devices, Portable Music and Data Processing Devices, and Tablet Computers*, No. 337-TA-794, U.S. International Trade Commission, July 5, 2013; Chetan Sharma, "Mobile Patents Landscape: An In-Depth Quantitative Analysis," Chetan Sharma Consulting, 2013.

274. **A few months later:** "Judge Sets New Trial Dates in Apple vs. Samsung Patent Battle," Reuters, April 30, 2013; "Strong Demand for Smartphones and Heated Vendor Competition Characterize the Worldwide Mobile Phone Market at the End of 2012, IDC Says," IDC press release, January 24, 2013.

Chapter 17: Critical Mass

276. **Outside the courts:** "Obama's Full 2013 State of the Union Address," YouTube video, uploaded by WSJDigitalNetwork, February 12, 2013; John Nosta, "With Apple CEO in Attendance, Obama Addresses Jobs & Wages," CNBC Storify page, http://storify.com/CNBC/with-apple-ceo-in-attendance-obama-to-talk-jobs-in; Josh Tyrangiel, "Tim Cook's Freshman Year: The Apple CEO Speaks," *BloombergBusinessweek*, December 6, 2012.

278. **Despite the president's stamp of approval:** Interview with David Wolf; Laurie Burkitt, "China Broadcaster Shows Up Apple, VW," *Wall Street Journal*, March 15, 2013; Philip Elmer-DeWitt, "Chinese TV Caught Asking Celebrities to Bash Apple," *Fortune*, March 16, 2013; CCTV Channel 2 segment, uploaded by NetEase, trans. by Violet Tian, http://news.163.com/13/0329/20/8R5LD72S00011229.html; Laure He, "Tim Cook's Apology Letter to Apple Customers in China," *Forbes*, April 3, 2013; Melanie Lee, "Sorry Apple Gets Respect in China After Tabloid Trial," Reuters, April 2, 2013; Li Hongbing, "Da Diao Ping Guo Wu Yu Lun Bi De Jiao Ao" [Defeat Apple's "Incomparable" Arrogance], trans. by Violet Tian, *People's Daily*, March 27, 2013.

283. **Nowhere was Apple's decline:** Jason Gilbert, "The Last 6 Times Tim Cook Has Talked, Apple's Stock Has Dropped," *Huffington Post*, February 28, 2013; Phil Goldstein, "Apple's Cook: Smartphone Market Is 'a Wide-Open Field,'" *FierceWireless*, February 12, 2013.

286. **But his oft-repeated line:** Interviews with people familiar with the situation, including Greg Dudey and other current and former Apple employees; Brad Stone, Adam Satariano, and Peter Burrows, "Mapping a Path Out of Steve Jobs's Shadow," *BloombergBusinessweek*, October 3, 2012; Jonathan Ive photos, Getty Images, http://www.gettyimages.com/editorial/jonathan-ive-pictures; Josh Tyrangiel, "Tim Cook's Freshman Year: The Apple CEO Speaks," *BloombergBusinessweek*, December 6, 2012.

289.   **On a cold afternoon in March:** Interview with person familiar with *Wall Street Journal* phone meeting with Schiller; "Samsung Unpacked 2013: Samsung Galaxy S 4 Launch," YouTube video, uploaded by "lifechannelable," March 14, 2013; Ina Fried and Lauren Goode, "Samsung Takes Its Flagship Smartphone into a New Galaxy," All Things Digital (blog), March 14, 2013, http://allthingsd.com/20130314/samsung-launches-galaxy-s-4/.

Chapter 18: Holy Grail

292.   **In the end:** "Samsung Unpacked 2013: Samsung Galaxy S4 Launch," YouTube video, uploaded by "lifechannelable," March 14, 2013; Molly Wood, "Samsung GS4 Launch: Tone-Deaf and Shockingly Sexist," CNET, March 14, 2013; Nilay Patel, "Samsung Weird: How a Phone Launch Went from Broadway Glitz to Sexist Mess," *Verge*, March 18, 2013.

293.   **Apple's woes were deepening:** Charles Duhigg and Keith Bradsher, "How the U.S. Lost Out on iPhone Work," *New York Times*, January 21, 2012; Charles Duhigg and David Kocieniewski, "How Apple Sidesteps Billions in Taxes," *New York Times*, April 28, 2012; Tyrangiel, "Tim Cook's Freshman Year"; "Testimony of Apple Inc. Before the Permanent Subcommittee on Investigations U.S. Senate," Apple, May 21, 2013; Permanent Subcommittee on Investigations, S. Rep, Memorandum on Offshore Profit Shifting and the U.S. Tax Code—Part 2 (Apple Inc.), May 21, 2013; Anna Palmer, "Apple Prepares for Washington Onslaught," *Politico*, May 20, 2013; Cecilia Kang, "Apple CEO Tim Cook to Propose Tax Overhaul," *Washington Post*, May 16, 2013; "Subcommittee to Examine Offshore Profit Shifting and Tax Avoidance by Apple Inc.," Carl Levin press release, May 20, 2013; data on lobbying spending by Center for Responsive Politics, http://www.apple.com/pr/pdf/Apple_Testimony_to_PSI.pdf.

297.   **On the morning of May 21, 2013:** "U.S. Tax Code and Offshore Holdings," C-SPAN video library, May 21, 2013; "*The New Yorker*'s George Packer on America's Unwinding Equality," *Forum with Michael Krasny*, KQED, June 6, 2013; George Packer, *The Unwinding: An Inner History of the New America* (New York: Farrar, Straus & Giroux, 2013).

298.   **"To whom much is given, much is required":** Though the inspiration for this quote likely came from the Bible in Luke 12:48 ("For unto whomsoever much is given, of him shall be much required"), Cook attributes it to John F. Kennedy. The president's exact words were, "For of those to whom much is given, much is required."

303.   **Within a few days, Apple:** *United States v. Apple Inc., et al.*, No. 12-cv-02826, Opinion & Order, September 6, 2012; *United States v. Apple Inc., et al.*, No. 12-cv-02826, complaint, April 11, 2012; *United States v. Apple Inc., et al.*, No. 12-cv-02826, hearing transcript, May 23, 2013.

Chapter 19: The Red Chair

306.   **In late May 2013, on the opening night:** Tim Cook, interview by Walt Mossberg and Kara Swisher, D11—All Things Digital conference, May 28, 2013, http://allthingsd.com/video/?video_id=0A0DDC54-6929-43AA-818E-3058B33077B7; Tim Cook, interview by Walt Mossberg and Kara Swisher, D10—All Things Digital conference, May 29, 2012, http://allthingsd.com/20120611/apples-tim-cook-says-hello-the-full-d10-interview-video/.

307.   **"You're being beaten up":** Mossberg's quote has been edited for clarity. His exact comment: "You're being beaten up by various governments over various things,

and I think worst of all, there's a, and your stock is down. And there's a sense—down significantly—and there's a sense that, I—actually I'll tell you something. I honestly don't even know the price of the stock but whatever. There's a sense that you may have lost your cool, that somebody else has got the cool. That Samsung has got the cool."

310.   **One of the toughest questions:** Tim Cook, interview by Walt Mossberg and Kara Swisher, D11—All Things Digital conference, May 28, 2013; *In the Matter of Certain Electronic Devices Including Wireless Communication Devices, Portable Music and Data Processing Devices, and Tablet Computers*, No. 337-TA-794, U.S. International Trade Commission, July 5, 2013, 58–60; *In the Matter of Certain Electronic Devices Including Wireless Communication Devices, Portable Music and Data Processing Devices, and Tablet Computers*, No. 337-TA-794, U.S. International Trade Commission, March 19, 2003, 18–19; Dan Levine, "Judge Who Shelved Apple Trial Says Patent System out of Sync," Reuters, July 5, 2012.

312.   **On the morning of June 3, 2013:** Interview with *Fortune* columnist Philip Elmer-DeWitt; *United States v. Apple Inc., et al.*, No. 12-cv-02826, transcripts, June 3–20, 2013; Philip Elmer-DeWitt, "The Apple e-Book Trial: The View from the Hard Benches," *Fortune*, July 11, 2013.

315.   **Almost three weeks later:** Interview with Philip Elmer-DeWitt; *United States v. Apple Inc., et al.*, No. 12-cv-02826, opinion and order, July 10, 2013.

Chapter 20: Manifesto

319.   **On a cloudy morning:** Interview with Jordan Eskenazi and other WWDC attendees.

321.   **When Tim Cook and his team:** "Apple—WWDC 2013," YouTube video, uploaded by Apple, June 19, 2013, http://www.youtube.com/watch?v=SRmjUzcpLO0; "Apple's WWDC 2013 Keynote: iOS 7, iTunes Radio And New MacBook Airs," ReadWrite (live blog), June 7, 2013, http://readwrite.com/2013/06/07/apple-wwdc-2013-keynote-live-coverage#awesm=~ofmSoX9lVCF7SG.

323.   **"In the phone market they are":** Matthias Hohensee, "Wir wollten die Angst vor Computern nehman," *Wirtschaftswoche*, February 7, 2013. This version was shared with the author by the article's writer, Matthias Hohensee. It is Wozniak's exact quote as was actually spoken by him in English. There are other versions attributed to Wozniak on the Internet, but those are English translations of the German translation.

326.   **The final disappointment:** Interview with Ken Segall; "Apple—WWDC 2013," YouTube video, uploaded by Apple, June 19, 2013, http://www.youtube.com/watch?v=S-RmjUzcpLO0; Peter Burrows, "Apple's TV Ads Touting Company Values Flop with Viewers," *Bloomberg*, June 27, 2013; Tim Nudd, "Ad of the Day: Apple," *Adweek*, June 11, 2013.

327.   **Ace Metrix:** After the Ace Metrix survey was released, some media questioned the accuracy of the data because it gave higher scores to Samsung, a client. The author determined that the results are still valid. Many companies retain the services of third-party research firms, and that doesn't automatically render the data to be questionable. Ace Metrix is a respectable firm that is often quoted by industry trade publications including *Adweek* and *Advertising Age*. Apple's former copywriter Ken Segall also validates Ace Metrix's findings on his blog, writing that the humorous in-your-face ads like Samsung's tend to do better than quiet ads like Apple's.

328.   **In the months that followed:** John Paczkowski, "The Incredible Shrinking Apple E-Book Remedy," *All Things Digital*, August, 28, 2013, http://allthingsd.com/20130828/

the-incredible-shrinking-apple-e-book-remedy/; Nate Raymond," U.S. Judge Wants External Monitor for Apple in E-books Case," Reuters, August 27, 2013, http://www.reuters.com/article/2013/08/27/net-us-apple-ebooks-idUSBRE97M0OU20130827; David McKenzie and Charles Riley, "Xiaomi CEO Tired of Steve Jobs Comparison," CNNMoney, September 13, 2013, http://money.cnn.com/2013/09/12/technology/xiaomi-phones-china/index.html; David Barboza, "In China, an Empire Built by Aping Apple," *New York Times*, June 4, 2013; Eric Pfanner, "Samsung's Profit Rises, but So Does the Competition," *New York Times*, July 26, 2013, http://www.nytimes.com/2013/07/26/technology/samsungs-profit-rises-but-so-does-the-competition.html; "Apple Unveils Panicked Man with No Ideas," *Onion*, September 20, 2013; "Apple Cedes Market Share in Smartphone Operating System Market as Android Surges and Windows Phone Gains, According to IDC," IDC, August 7, 2013; "Growth Accelerates in the Worldwide Mobile Phone and Smartphone Markets in the Second Quarter, According to IDC," IDC, July 25, 2013; Sam Grobart, "Apple Chiefs Discuss Strategy, Market Share—and the New iPhones," *BloombergBusinessweek*, September 19, 2013; Harry McCracken and Lev Grossman, "Google vs. Death: How CEO Larry Page Has Transformed the Search Giant into a Factory for Moonshots," *Time*, September 30, 2013; Marco Arment, "Off," Marco.org, October 24, 2013.

Epilogue

330. **From the start, Apple embodied:** Steve Jobs himself referred to Apple's exceptionalism at a press conference on July 17, 2010, to address the iPhone 4's antenna problems. He said the following, according to Reuters. "I guess it's just human nature that if someone or some organization gets really successful there's just a group of people who want to tear it down. I see it happening with Google."

"I see some of these people jumping on us now. It's like I am not sure what you are after here. Would you rather we were a Korean company instead of an American company? You do not like the fact that we are innovating right here in America and leading the world in what we do? . . . . Of course we're human, of course we make mistakes."

Index

Maghsoodloo, Saeed, 98
Ma Jun, 119–21, 122–23, 126
Malkovich, John, 153
management (upper management), 37,
57–63, 67, 99–100, 133–34; Amelio as
CEO, 12; board of directors, 28–29,
277; "Bottom 100," 66–67; compensation, 63, 64; Cook as CEO, 91,
93, 104–7, 126–28, 130–32, 140–41,
156–59, 185, 194, 195, 219, 241–49,
278–91, 294–303, 306–10, 319–27,
336–38; Cook as COO, 62; Cook gives
stock bonuses, 141; Cook's consolida-
tion of power, 61–62; Cook's firings,
244–45, 248; Cook's performance,
36–38, 51, 55–56, 58–62; IBM alum,
56, 61, 123; Jobs as CEO, 12–15, 39,
80–84, 99, 105, 130–31, 246, 249,
287; Jobs's illness and, 16, 24, 28–29;
Jobs's lieutenants, 14, 67, 106, 132–40,
320; Jobs's medical leaves and, 27, 31,
36–38; kickback scandal, 63–64; loss
of intensity, 287–88; meritocracy and,
13; Oppenheimer as CFO, 20, 158, 300;
personalities of major figures, 132–41;
restructuring (1998), 57; rivalries,
288; Sculley as CEO, 11, 188–89, 194;
succession plan, 28, 82, 83; Top 100
meeting and, 66–71. *See also* Browett,
John; Cook, Tim; Forstall, Scott; Ive,
Jonathan; Schiller, Phil
Mavericks operating system, 324
Ma Xiangqian, 64, 111
Mays, Ray, 56
McCain, John, 297, 299–300, 301
McCartney, Paul, 53, 130, 131
McCluney, Jim, 57
McElhinny, Harold, 216–17, 218, 227, 271
McKenna, Regis, 243
Methodist University Hospital, 33–34
Microsoft, 11, 85, 171, 295; as Apple
competitor, 11, 188, 249; Nokia and,
177–78
Miller, Andy, 69
Milunovich, Steve, 234
MobileMe, 156
mobile phones, 247–48; China market for,
332; Chinese companies and, 282, 332;
disruption theory and, 185; market
share, 186, 210, 239, 264, 278, 285;
Samsung patents, 274. *See also* Apple

v. Samsung; Google; HTC; iPhone;
Nokia; Samsung
Morita, Akio, 192, 194
Morrison Foerster law firm, 216, 233
Mossberg, Walt, 34–35; All Things Digital
conference (2013), 306, 307, 308;
antitrust case against Apple and, 313;
piece on Jobs prior to his death, 83;
review of iPad, 44
Motorola, 60, 91, 119, 149, 167, 168, 176,
190, 200; Apple lawsuit, 172, 181,
310–12; *Apple v. Samsung* and, 210;
Razr phone, 173
Mueller, Florian, 211–12, 230–31
Mukunda, Gautam, 246; business physics,
248, 287; leadership analysis, 246–48,
334
Munster, Gene, 28, 38, 153
Murdoch, Rupert, 42, 89, 163
Murphy, Charles, 97
Musika, Terry L., 182

Natural Resources Defense Council
(NRDC), 123
NewCastle Polytechnic, 137
Newton PDA, 188–89
New York Times: article on Jobs's health,
20; exposé on Apple's tax avoid-
ance, 294; iEconomy series on Apple,
125–26, 197, 293; Jobs pitches iPad
to, 41–42; Nocera column, 20, 23;
report on iPad manufacturing labor
conditions and practices, 126; review
of iPad, 44
NeXT, 12, 21, 75, 134, 156
Nike, 91
Nocera, Joe, 20–22, 23
Nokia, 119, 148, 165–66, 200, 274; Apple
lawsuit against, 172, 177; Windows
Mobile, 177–78
Nudd, Tim, 327

Obama, Barack, 276–77, 312
Obama, Michelle, 276, 277
O'Brien, Conan, 223–24
Odle, Sheila, 61
O'Melveny & Myers law firm, 295–96
Onion, 333
Oppenheimer, Peter, 20, 36, 158, 300, 301
Ornish, Dean, 84
O'Sullivan, Joe, 57–58, 102

About the Author

Yukari Iwatani Kane is a former technology reporter for the *Wall Street Journal* in San Francisco who has covered Apple and related technology trends. In 2011, she was named a Gerald Loeb Award finalist as part of a *Wall Street Journal* team for a series on Internet privacy. Prior to covering Apple, she was based in Tokyo, reporting on the Japanese technology industry, including Sony and Nintendo. Before working at the *Wall Street Journal*, she spent seven years writing technology stories for Reuters in Tokyo, Chicago, San Francisco, and Washington, D.C.